Studies in Computational Intelligence

Volume 714

Series editor

Janusz Kacprzyk, Polish Academy of Sciences, Warsaw, Poland
e-mail: kacprzyk@ibspan.waw.pl

About this Series

The series "Studies in Computational Intelligence" (SCI) publishes new developments and advances in the various areas of computational intelligence—quickly and with a high quality. The intent is to cover the theory, applications, and design methods of computational intelligence, as embedded in the fields of engineering, computer science, physics and life sciences, as well as the methodologies behind them. The series contains monographs, lecture notes and edited volumes in computational intelligence spanning the areas of neural networks, connectionist systems, genetic algorithms, evolutionary computation, artificial intelligence, cellular automata, self-organizing systems, soft computing, fuzzy systems, and hybrid intelligent systems. Of particular value to both the contributors and the readership are the short publication timeframe and the worldwide distribution, which enable both wide and rapid dissemination of research output.

More information about this series at http://www.springer.com/series/7092

Alexander P. Sukhodolov
Elena G. Popkova · Irina M. Kuzlaeva

Internet Economy vs Classic Economy: Struggle of Contradictions

Alexander P. Sukhodolov
Baikal State University
Irkutsk
Russia

Irina M. Kuzlaeva
Volgograd State Technical University
Volgograd
Russia

Elena G. Popkova
Volgograd State Technical University
Volgograd
Russia

ISSN 1860-949X ISSN 1860-9503 (electronic)
Studies in Computational Intelligence
ISBN 978-3-319-86822-6 ISBN 978-3-319-60273-8 (eBook)
DOI 10.1007/978-3-319-60273-8

Printed on acid-free paper

This Springer imprint is published by Springer Nature
The registered company is Springer International Publishing AG
The registered company address is: Gewerbestrasse 11, 6330 Cham, Switzerland

Contents

Introduction

Internet economy is a new level of development of economic science. This monograph studies its sense and basic principles, determines the factors influencing its functioning and development, analyzes the crisis of Internet economy, and performs comparative analysis of Internet economy and classic economy.

In the 21st century, classic economy was replaced by Internet economy which supposes a completely new approach to doing business according to the new technological mode, changes relations between sellers and buyers in the market, and requires reconsideration of one of the most important economic categories—product.

Thus, there arises an objective necessity for scientific study of this new economic reality, which is done in this monograph. It seeks the goal of comparing Internet economy to classic economy, determining their similarities and differences, pros and cons, and determining possibilities and perspectives of building of Internet economy in modern Russia. Internet economy is a new level of development of economic science.

The monograph studies its sense and basic principles, determines the factors influencing its functioning and development, analyzes the crisis of Internet economy, and performs comparative analysis of Internet economy and classic economy. A key peculiarity of this book is the problem approach to study of Internet economy consisting in its opposing to classic economy, complex analysis, and description of experience of building Internet economy in modern Russia. The targeted audience of this monograph includes postgraduates and lecturers of higher educational establishments and researchers who study foundations of modern macro-economics.

Chapter 1
Production and Economic Relations on the Internet: Another Level of Development of Economic Science

1.1 Unconventional Approach to Treatment of the New Economy

At present, it is easier to acknowledge that the issue of active involvement of high information technologies into various spheres of modern society's life is rarely discussed in scientific and civilian groups. Discussions regarding the material side of this problem—say, formation of natural and side effect received from mutual intertwining of the global information and communication environment and entrepreneurial capabilities seem to be very interesting. Such popular tandem became manifested in appearance of Internet economy—economy that develops according to different laws and categories that became such due to peculiarities of the environment of their appearance and functioning—the Internet.

Very often, the use of possibilities of the World Wide Web by average users is brought down to communication in social networks and forums, personal blogs, and various e-mails. Those who want to use their personal time with maximal effectiveness, see the Internet as a platform for search for certain goods, comparing prices and reviews, and, finally, purchases with simple electronic operations. Unlike individuals, legal entities use the services of Internet providers for the purpose of constant communication with suppliers, partnership and customer groups. For a state, the Internet is a powerful and universal tool that provides a large specter of various functions and possibilities: from realization within quick and convenient administering and provision of services to certain people and companies to popularization of various laws.

As a result of the performed research of a wide range of treatments of the general notion and separate elements of category of "Internet economy", that take place in actual economic works, the authors determined the necessity to inform the readers of the idea that the limits of defining the viewed notion have remained obscure for a long time bereaving the interested people of the possibility to form a clear notion of its as of a specific direction of modern economic theory. Thus it became a habit

A.P. Sukhodolov et al., *Internet Economy vs Classic Economy: Struggle of Contradictions*, Studies in Computational Intelligence 714,
DOI 10.1007/978-3-319-60273-8_1

that people usually use the notions "network economy", "virtual economy" and "E-economy".

This aspect should be viewed in detail. As a matter of fact, the notion "Internet economy" could be compared to "network economy" and "virtual economy". The category "E-economy" is more complicated, as it reflects a wider circle of the issues of functioning of economic nature that the problems related to the Internet, its means, technological tools and applications. Thus, "E-economy" is a system of economic relations that pierce all stages of commodity and service production, distribution, movement, and realization of various goods and that develop against the background of electronic exchange of data through electronic networks.

It seems that foundations of E-economy consist of information technologies, the problem of evolution of which—due to its topicality—is reflected in multiple works of technical and humanitarian nature. At present, information technologies are computerized ways of creation, storing, transfer, and processing of data in the form of knowledge received by scientific methods and methods of their application. The phenomenon of computerization supposes replacement of manual and mechanical processing of data by electronic. It enters our life so confidently that the object of this phenomenon are information flows belonging to production and non-production activities.

The earliest elements of E-commerce, as a form of electronic entrepreneurship, appeared in 1970s. Back them, they were used by transport organizations for the purpose of provision of tickets reservation services. A substantial drawback of early forms of E-business was their limited accessibility. Invention of the Internet stimulated appearance of a large number of new forms of E-entrepreneurship and commerce—in particular, Internet business and Internet commerce—which was a sign of establishment of so called Internet economy that was seen as a systemically organized and multi-level structure founding on the relations between economic actors in the global computer network.

In a wide sense, Internet economy is a new stage of perfection of production forces, accompanied by growth of the share of Internet products and services in GDP of modern countries and appearance of international economic Internet space that ensure efficient interaction of individuals involved in the process of production, realization, and consumption of Internet products.

Year 1983 is considered to be the initial point in the process of development of online economy—it ended the period of integration of specific local networks and final establishment of the common space for moving the information and communication flows—"Internet", the name for which was borrowed from the data transfer protocol.

Thus, online economy became a result of quick development of E-economy: structural parts of the former continue and complete the whole totality of the latter, forming a single totality, and, therefore, creating the economy of innovational and information type as a new level of development of post-industrial concept, the author of which is Bell (1973).

This scheme helps to give the readers a vivid idea of mutual existence of the processes that take place in information economy. Internet technologies, as natural consequences of improvement of information technologies, absorb all components of the system of online economy. This system is reflected in four key elements: online business, online commerce, online project, and company's web-site.

In its turn, online business is one of the results of improvement of E-business, which has been a sub-system of E-economy, acting on the basis of information technologies. Online commerce is a separate branch of E-commerce that became independent in early 1990s. The foundations of online commerce consist of online structures of information management, an important part of which is online project, which is one of the links of online economy. In a classic form, online project is a wide specter of many-profile documents that contain detailed description of certain measures on creation, exploitation, and provision of effective functioning of the project within the Internet. A peculiar feature that distinguishes Internet economy from E-economy is the formulating basis—web-site of the Internet economy. Electronic economy acknowledged another foundation of its existence—servers, data bases, and program codes that are important components in the processes of development and servicing web-sites. Under the present conditions of development of world economic environment, web-sites contain all information data that have to be transferred to the potential Internet client.

Sense and specifics of Internet economy are studied in many works of modern authors, results of which are published in the peer-reviewed international scientific journals and books. Some authors define Internet economy as a completely new type of economic system that possesses unique (revolutionary) specific characteristics that cannot be formed on the basis of other such types of systems.

Within this approach, Gnezdova et al. (2016) see Internet economy as economic system that generates and actively implements Internet technologies into all spheres. Chen (2016) distinguished such important features of Internet economy as wide distribution and use of the Internet and its key role in creation of added value. Choi et al. (2014) pay attention to reformation of economic relations in view of Internet-technologies, as the most important characteristic of Internet economy.

Other researchers think that Internet economy is a new stage of evolution of economic systems that follows the post-industrial economy (or knowledge economy that is sometimes distinguished as a latter system). Thus, Lavrinenko and Okhotina (2015) state that Internet economy cannot be formed on the bass of industrial or pre-industrial economy, for it cannot "skip" one or several levels in economic development of economic systems.

Taylor et al. (2012) pay attention to the fact that service sphere that plays an important role in formation of Internet economy reaches the necessary scale and level of development in post-industrial economy. Yadav et al. (2015) come to the conclusion that Internet economy is not the highest level of development of economic system but another stage followed by others—but it is not possible to forecast these new stages due to obscure course of future technologies' development.

Other scholars state that Internet economy is just a concept, a new possibility that appeared with emergence and distribution of Internet technologies and that is not yet realizes and perhaps will not be realized in the real world. Followers of this approach—Jarrett and Wittkower (2016)—state that Internet economy cannot be

Table 1.1 Results of comparative analysis of approaches to treatment of Internet economy

Approach	Sense of approach	Representatives of the approach
Revolutionary	Completely new type of economic system that possesses unique (revolutionary) specific characteristics	Chen, Choi, Williams, Ha
Evolutionary	New stage of evolution of economic systems that follows post-industrial economy	Lavrinenko, Okhotina, Yadav, Chauhan, Pathak
Conceptual	Concept, new possibility that appeared with emergence and distribution of Internet technologies that is not yet realized	Jarrett, Wittkower, Fuchs, Bolin

formed in practice due to lack of technical possibility and society's reluctance to fully refuse from classic economic deals and traditional form of entrepreneurship.

Fuchs (2009) writes that Internet economy is a myth created by scholars to denote a direction of development of world economic system and quicken its overcoming of the global crisis. Fuchs and Bolin (2012) state that Internet economy could be created inly in "ideal" conditions and, like purely market economy, is not a goal for real economic systems that can only approach its formation (Table 1.1).

The idea of Internet economy is so popular that this issue is viewed in many books. For example, the book by Powers and Jablonski (2015) contains the result of theoretical consideration of the sense of Internet economy. Birner and Garrouste (2013) consider perspectives of development of Internet economy in Austria, and (OECD 2013) views the rise of Internet economy and its establishment in modern economic systems.

1.2 Factors of Functioning and Development of Internet Economy

Good stability of rates of development of American economic space in 1990s, which was accompanies by low inflation level and low unemployment level, as well as quick evolution of the information sector together with powerful growth of prices of stock of companies working in its perimeter, made many people think that improvement of information technologies influences the vector of transformation of macro-economic regularities that earlier seemed constant.

These beliefs were not shattered even by the '97 Asian crisis that led to downfall of stock prices of a lot of companies, including of American descent, that were involved in production and realization of high technologies. Quick decrease of the above indicators forestalled their quick rehabilitation and further growth. Thus, late 1990s showed new growth of rates of development of the U.S. economy. Pitifully, the possibilities of economic science didn't allow the scholars to forecast or even guess the probability of these events or explain the nature of their emergence.

The hypothesis that described structural transformation in American economic system, which took place as a result of quick development of IT potential of the country, was called a new economic paradigm. The role of new variables that determined structural features of a certain economic system, belongs in the new paradigm to the natural unemployment level and stable rates of economic growth in the conditions of low inflation. In other worlds, the key suppositions that constitute this paradigm could be presented in the following way: the rates of economic growth increased a lot, with the relatively constant level of unemployment and inflation; natural unemployment level decreased as well. Thus, the seemingly universal hypothesis on negative character of relations between inflation and unemployment lost its topicality.

Despite the fact that economy of post-industrial type is easily observed in the part that is directly related to computer technologies, in particular, the Internet, and the indicators characterizing its development are easily measured, the new paradigm became a conductor, the input of which included highly-efficient calculation machines and output—low level of unemployment and stable growth without inflation. Unfortunately, the paradigm cannot provide their detailed and qualitative description. In view of the fact that the theory of general macro-economic balance does not have analogs that could be applied to Internet economy, it is possible to note that models application of which enables to describe specific moments of existence of such balance have already appeared. These include economic models of network interactions between buyers and sellers via the Internet, etc.

Internet space, as a core of Internet economy, directly and indirectly influence the whole totality of landmarks of activities of online business, which include commerce, stock market, government, insurance, consulting, mass media, investing, banking, education, tourism, hosting, money, providing, freelance, logistics, etc.

Direct factors that influence the course of development of Internet economy include the following:

- level of paying capacity of citizens;
- computer literacy of citizens;
- geo-natural capabilities of the territory;
- internal and external political situation;
- level of development of information and communication technologies at the state level.

The group of direct factors that influence Internet economy include the following:

- activity of Internet audience;
- level of state preparation in the sphere of legal acts that regulate business relations on the Internet;

- level of development of large and small organizations that realizes their activities with the use of information and communication technologies;
- general level of development of Internet infrastructure, existence and improvement of which allows achieving reduction of distance within the Internet between regular users and business actors.

There is opinion that at this historical stage developed countries are peculiar for active growth of labor efficiency that depends on the level of development of information technologies. However, there are other positions as well. Thus, certain scholars think that the results of analysis of empirical data by the example of American economy allow stating the absence of long-term increase of growth rates of labor efficiency and related economic growth. All suppositions on transformation of the character of connection between unemployment and inflation also contradict the main postulates of the traditional economy. Despite this, there are those who do not argue the information technologies capabilities to influence macro-economic variables.

In our opinion, increase of growth rates of American economy in early 21st century is only partially connected to increase of labor efficiency in information sector of economy. This process was influenced by the phenomenon of unequal exchange between two sub-levels of public reproduction, when the cheaper volume of benefits of information character is exchanged for the larger volume of usual products. In this case, it is possible to observe growth of prices for goods of information sector of economy, which leads to increase of estimate indicators of labor efficiency and production effectiveness. At the first stages of development of economy of information type, specific rent profit appears, which is substantiated by comparative rarity of information products. All the above becomes a reason for formation of inequality of not only the cost proportions but norms of entrepreneurial income, as well as payments for labor employees at both sub-levels.

The phenomenon of non-equivalent exchange could be seen not only in reproduction structure of the specific state but in the world economy on the whole. The countries that entered the age of information technologies using the non-equivalent exchange within their foreign trade activities form the system of exploitation of states with developed economy, getting rare material and intellectual resources for free in exchange for cheap copies of information products. As a rule, the money received from such operations is spent for purchase and placement of material production objects in developing countries. Thus, there is formation of a vicious circle of constant lagging of the countries of the third word, accompanies by strengthening of the structure of their resource specialization and intellectual "slavery".

The condition, under which the similar actions of state with developed economy, as well as transnational corporations that sell information products, lead to certain reduction of the speed of scientific and technical and economic progress of society is not to be wondered at; this is caused by their striving to preserve and support the achieved status quo and monopolistic power over the objects of intellectual property. In this situation, so called "pirates" do a good cause, competing with information tycoons that put effective realization of the procedure of free distribution of information in dependence on their huge profits.

At present, Internet is a tool for realization of free information exchanges at the global scale, which influences the democratization of intellectual market. Let us supposes that the future's existence is based not on large corporations but thousands of independent individuals who create products of intellectual labor, demanded by the society, and who enjoy it. At that, we do not refuse the possibility of them receiving decent money for their labor activities.

Increase of the volume of total public product, according to economic laws, should be accompanied by growth of capital of two reproduction sub-levels according to cost and natural and material content. The law of quick development of information sector of economy, being a base of expanded reproduction in the period of emergence and improvement of society of the post-industrial type, interacts with the procedure of capital accumulation.

Speaking of the category of capital, one must say that the process of evolution of information economy cannot exist without its transfer between separate spheres or states. Internet here plays his own role: effective existence of the world market capital that includes stock market, currency market, and inter-bank deals is impossible without it.

Let us conclude: improvement of the global network Internet should become one of the high-priority strategic directions of development of any country that strived for taking a decent place in the global arena. As of now, certain states actively develop and realize certain elements and directions of a complex strategy of development of their economic systems, successfully applying Internet technologies.

1.3 Comparative Analysis of Internet Economy and Classic Economy

One should understand that quick evolution of Internet economy in all its directions provokes various changes in the global practice of economic development and leads to change of classic postulates of economic theory. Thus, the peculiarities of Internet economy that stimulate such metamorphoses include the following:

- structure of network economy expenses (practical striving of marginal costs of expansion of production to zero) that hinders many competing subjects in the perimeter of one market segment. As a rule, only the strongest survive, as consumer expenses aimed at supplier replacement are very close to zero;
- information components that more or less accounts for cost of good or service;
- intensive use of electronic payments systems during realization of a wide specter of online transactions within the scale of economic space of a specific country leads to significant increase of the level of total income;

- value of the results of labor that comes from the fact of their multitude. Indeed, the tendency of appearance of additional objects in Internet space shows real growth of positive influence of this factor on all of its consumer audience. Such peculiarity of Internet economy directly contradicts two fundamental laws of traditional economic theory: value is created by rarity of resources, and excess of things brings them down to minimum;
- exponential dependence between growth of the value of participation in online economy and growth of the number of Internet users;
- low level of constant expenses, marginal costs, and quick distribution of products, peculiar for the analyzed type of economy, stimulate reduction of time interval required for reaching the stage of quick growth in economy of a classic type;
- Internet economy founding on the law of network action—the law of growing feedback. It should be noted that as opposed to industrial economy, within which growth of feedback is a results of unification of common efforts of various organizations, which, actually, should receive the results of this process, increase of feedback in a network is done by the whole network and is distributed by all active subjects;
- Internet economy's capability to provide opportunities for creative manifestation of a personality within personality's functioning: if combination of copying and automatization is a quickly depreciating phenomenon, the tandem of originality and creativity is something that begins to develop.

Thus, we have a chance to witness realization of very interesting processes that are brought down to formation of a new direction in economic science—Internet economy, which surely will lead to cardinal changes of the existing stereotypes in the sphere of search, systematization, and application of information data. We live in the age of informatization, computerization, and Internetization of most countries of the world. This will be followed by appearance of new factors of economic growth and development of information and innovational type of reproduction, which makes quick perfection of Internet technologies very real (Maksiyanova 2013).

Up until now, the issue of equivalence/non-equivalence of exchange processes in Internet economy is very topical. Modern economists are very interested in it—especially now, in the conditions of global change of the traditional economic mode. The traditional structure is industrial structure of economy, within which the elements peculiar for archaic economic systems can exist. Economy of the new time finds itself at the information stage of social development.

The subjects acting in a traditional economy seek the goal of maximization of profit in the conditions of limited material and labor resources. Of course, such system must have a market economic mechanism. Usually, equivalence of exchange in market economy is explained by one of two main methods: based on the theory of limited usefulness or the labor theory of cost. Thus, exchange is equivalent when there is equality between values of limited usefulness of the exchanged products or the same volume of socially necessary labor resources spent for their production.

The basis of resource provision in a new economy consists of knowledge and information which are actually unlimited. Drawing a parallel between the value of material bearers of information data and expenses oriented at their creation, it is possible to conclude that the first component is more than low. A final Internet product was, is, and will be a result of a unique creative process. However, distribution of its multiple copies is usually very cheap. It seems to be causing such phenomenon as non-equivalent exchange.

In our understanding, violation of equivalence of exchange procedures in the economy of a new type is absent. Still there is a question: what is equivalent? While in classic economy a general measure is money, something else performs its role in information economy. In order to have no doubts as to this issue, it is expedient to analyze the process of exchange of a commodity unit received as a result of application of limited resources of traditional economy for an information product of the new economic mode. It is also necessary to consider exchange of creative goods in the new economy.

Let us view the first situation. Let us imagine that an intellectual product is to be sold for a price of a cheap commodity, and the manufacturer receives only a small share of his creative labor expenses. Modeling of the procedure of realization of the whole set of material commodities will helps us to see the presence of equivalence in the phenomenon that interests us. The sense of this equivalence consists in the idea of limitation of intellectual capabilities as a resource in the traditional economy, due to which the income of realized products' owner has a rent character.

Manifestation of popularization of contra fact information goods is also present on our scheme, as it leads to growth of the level of popularity of the same manufacturer—"pirate" copying includes a hidden advertising aspect. There is also moral ageing of information products. Even if their social value as the value of components of humankind's culture can stay at a high level, it is necessary to acknowledge that money evaluation can sometimes drop to zero, but usually it becomes a result of constant changes in various theoretical and practical directions.

A case of exchange of information goods in the conditions of functioning of new economy that somewhat looks like exchange of gifts is very interesting—when, for example, subjects of creative activities place their unique materials on the Internet for free. From the positions of axiology, such exchange is equivalent regardless of real cost of objects provided for it. Certain groups of scholars support free distribution of musical and literature products in their totality online, which will surely lead to emergence of a new form of equivalence. It is necessary to mention the following condition: active popularization of counterfeit information products in the world created a powerful stimulus for development of the global economy. In this regard, activity of large-scale IT corporations restrains further distribution of business connections in the "new" economy.

Thus, equivalence of exchange is seen during exchanging processes between economies of the traditional and new types in monetary form, and in new economy—in non-monetary. It is necessary to note that money's preserving the role of equivalent is possible under the conditions when at least one party of the exchange is limited material and labor resources. Due to the fact that their unlimitedness seems to be irresistible, money will be present in our life for many a century, continuing to influence the economy. In our opinion, society will not be able to refuse from money in future, so discussion on "post-economic formation" as self-sufficient structure is senseless—at least, now. It has a certain sense for the theory of economy as a science and for economists: the former is not threatened with loss of actuality, and the latter are not threatened with loss of employment.

The sphere of application of the usual economic theory will be narrowing. Let us suppose that the common exchange equivalent in a "new" economic system will be a word, which is a certain set of sounds necessary for living of all humankind in a certain time period. Also, economy will be closer to other social sciences—e.g., philosophy and sociology.

It is necessary to acknowledge that certain aspects of modern reality go beyond the limits of existence of the institutional scheme of the Western science on society that studies the problem of improvement of online economy. Social science of the present should be based on the methodological principle of polyvariant ways of development. Of course, this path has a lot of doubtful turns and search for approaches to solving difficult tasks, the road of formation of a new language of science, new systems of effective knowledge, appearance of new objects of research that are reflected in totality of elements of rational treatment of the world—a decent alternative to modern science.

Chapter 2
Internet Economy: Existence from the Point of View of Micro-economic Aspect

2.1 Notion and Sense of the Category "Internet Product"

In our opinion, deeper understanding of the problem of presence of Internet economy in our life could be reached by viewing this phenomenon on micro-level, the simplest exchange item at which has been—from the very beginning of existence of economic relations on the Internet up until now—Internet product. It should be noted that the category "Internet product" consists of two important components: Internet service and Internet product.

Internet service, which supposes Internet provider's providing access to the network for one or another participant, traditionally possesses all "commodity" features, being a limited means—in one case, in the scale of space, in other—in time scale. Based on this, it is possible to say that most of postulated of the theory of limited usefulness and labor theory of cost can be applied to it.

The situation with Internet product is different, as for a consumer it is a result of intellectual and creative activities of individuals in the form fit for transfer on the Internet. Let us pay attention to the fact that Internet product, functioning of which is based on information, differs from information, as the contents of the latter is wider, for it can be taken from various sources. So, Internet product is a computer and information benefit, transferred from one basic station to another with the help of Internet.

Unlike traditional product, Internet product has its specificity expressed in non-material nature; low level of information carrier; connection to its source; unlimited character of distribution and application; exterritoriality; saturation and multiplicity of consumer characteristics; single nature of purchase and multiple nature of use; duration with relative aptitude to moral ageing of information.

Existence of Internet product is not always brought down to its commodity existence. All doubts regarding this issue could be dispersed by viewing the key attributes of Internet product as a commodity—its usefulness and value.

A.P. Sukhodolov et al., *Internet Economy vs Classic Economy: Struggle of Contradictions*, Studies in Computational Intelligence 714,
DOI 10.1007/978-3-319-60273-8_2

Usefulness—or, as it is called otherwise, consumer value—of Internet product is reflected in its having the role of a certain resources for various material and non-material objects. Apart from everything else, Internet product is present in the global information and communication environment as an element of the system of stock, commodity, and financial markets.

Consumer qualities of Internet product are not the full totality of information collected in various documents for the purpose of their further application within the search for solving specific problems. Indeed, it is a set of data presented in special formats the work with which allows a consumer to successfully solve the problems of economic, social, psychological, and other directions.

A quality of Internet product, as a commodity, to be viewed is its value characterizing its form under the conditions of commodity-money exchange. However, classic economic theories in their initial form do not always allow substantiating the proportions of commodity exchange in Internet economy.

Despite the above circumstance, commodity and money exchange in exists in Internet economy, as it is accompanied by constant development of real production sector—products in it are created by material resources of a limited character. Moreover, activity of Internet economy is built on the basis of limited labor capabilities—possibilities of a generator or, in other worlds, creator of actual information data. Such a unique resource seems nowadays to be a single case of existence of new value creator in information and communication space of the Internet.

As we see it, Internet product is something that is characterized by non-commodity contents and that manifests itself from the point of view of usefulness. Obviously, if need arises, its form can become commodity one, as at any moment there may arise a need for its exchange for product of commodity sector of economy, based on limited material resources and the law of reducing feedback of production means.

We know already: formation of the initial information material, as a basis of Internet product, takes place by means of processes of intellectual labor object that creates unique content. Indeed, the first copy of the product has high value, which can include separate quasi rents. However, even in such circumstances, Internet product may not be a commodity. For example, this may happen if any Internet use has access to information important for the public, posted by its creator on a free basis.

Product of information content acquires commodity form when the issue of its realization depends directly on the quality of artificial limitation of access to data constituting the its integrity, by means of various tools of online protection (PIN, license, etc.). All of this provokes a just growth of contradiction between non-commodity contents and commodity form of Internet product, which is based on emergence of another form of value of all following copies of information product that have authentic value.

The next issue to be considered is the issue of measuring and final considering the cost of Internet product. Some scholars, including Valtukh (2001), deem it expedient to measure the cost by the quantity of information data that are to be determined by their rarity.

As a matter of fact, probability that is a part of calculation of information cost of a certain objects as an important calculation component is the value reflecting the level of distribution of this object in the perimeter of a certain system. The offered proofs allow connecting cost not to information, materialized in contents and characteristics of the studied object, but to its popularity or rarity.

In our high-tech age, Internet product, having conquered its niche in the market space, entered the period of their presence in the information world when its cost is determined by rarity that has a load of social direction, as the set cost can manifest itself only in the moment of exchange of electronic product for other commodity items. It should be noted that the above relates only to the initial variant of existence of Internet product, its original form. Its second and further copies are far from the category "rarity"—that's why they crate only appearance based on monopolistic ownership of information benefit.

Based on the fact that the second and further copies of the original Internet product are not rare due to simplicity of their distribution, it is possible to state that Internet economy influences not only the law of cost but the law of reducing marginal usefulness. In the analyzed reality, its work looks differently: both marginal usefulness and value of Internet product, treated as addition of new useful information data that appear in the basis of new knowledge, increase with growth of the number of users who want to possess it.

It should be noted that each new Internet user influences the increase of usefulness of the known information and communication space for other participants. At that, the Internet benefits as well, for such changes lead to growth of its value. However, this process is peculiar for its non-linear character. This was a matter of interest for Metcalfe (2013), who invented localized network technology. In early 1970s, he came to the conclusion that in order to possess the network value it is necessary to achieve critical mass, but small networks of local character that together create one large network start to increase their value with large force.

Results of Metcalfe's observations became a cause for formulation of the law called after him—according to it, value of any network space is an equivalent of the square of connection nodes for its participant. Thus, usefulness from connection to network space grows by exponent due to increase of the number of its components. There are statements that this law decreases the real level of growth of network value, which is really larger.

The Metcalfe's observations consisted of the idea of existence of a telephone network which enabled to individuals to have connection, which meant the following: total volume of potential connections in this case had to depend on the object of paired connections in the network. However, existing networks allow for connection with three and more participants—so, their value grows much quicker.

Let us add: positive effects from functioning of network and non-linear growth of its value show themselves not from the beginning but from the moment when it reaches so called critical mass. In our opinion, this decisive moment already happened in Russian and foreign practice. The Internet is very popular with the inhabitants of the Earth.

Speaking of the problem of determining the value of Internet product, it is necessary to remember the issue of its life cycle. One should remember that duration of life cycle on the whole and its separate cycles depend on the commodity item and on market space within which it is realized. Raw materials are peculiar for longer life cycle, while final products are peculiar for short life-cycle. At last, high-tech products have a whole life span due to quick moral ageing.

Life cycle of information data as a commodity could vary due to various levels of actuality of their content in a wide range: from ages to several second. On the whole, the main difference between life cycle of Internet products and classic product is their duration that in the first case depends on the importance of information on the Internet. At the same time, there is a growing problem of contamination of Internet space with undesired commercial and non-commercial information that create too much load on channels of telecommunication connection. Such situation leads to aggravation of contradiction of information excess between the large volume of undesired, even virus, content and lack of necessary and useful information.

It is easy to conclude that information, which at present is a specific type of resources, production factor, and public improvement, is a special type of product that possesses relatively expressed commodity characteristics. Obviously, domestic and foreign information markets are multi-level and dynamic environments. This could mean only the following: intense use of leading technologies will allow them to differentiate by means of genesis of new needs of society (2).

2.2 The Nature of Mutual Functioning of Demand and Offer in the Market of Internet Goods

The fact that Internet economy became an actively developing sector of any self-sufficient country's national economy is not surprising—nor that the services provided within it acquire mass character.

In view of such circumstances, it is expedient to pay attention to one of the elementary types of economic interaction that take place on the Internet—namely, exchange of products and services from the point of view of the mechanism of demand and offer, price and competition, which form it.

As was mentioned, scenarios of economic relations that emerge on the Internet sometimes do not conform to traditional economic laws. This is really true: usual models of demand and offer that are fit for analysis and description of market mechanism of pricing in the market of standard economic benefits can be unacceptable for studying the market process that take place during exchange of information commodity items.

The classic economic theory is based on the principle of decreasing feedback of production means that manifests in the conditions of resource limitation. As a rule, it is not necessary to explain the sense of this statement to people with average level of economic knowledge. However, there could be differences regarding its universality in even in serious scientific circles. For example, representatives of neo-classic school refuse the applicability of this principle to conditions of modern development of economy. It is necessary to note that Marshall (1923), who set the foundation of the existing economic system, presented three possible states of production with three corresponding situations—situations of constant, growing, and decreasing feedback of production means.

In the situation of growing feedback, it is possible to see practical coincidence of the offer curve with the demand curve.

If the resources are unlimited, being valuable information, the balance point of demand and offer will strive for zero value of price. Such state of economic circumstances brings society to communist state that supposes that the main technological dominant is human with all his intellectual and spiritual capabilities.

As was mentioned above, in the process of improvement in the global scale the Internet network overcame a specific critical point, which allows stating that further interest will lie in the situation of emergence of demand for the first copy of Internet product, the price for which has reached its maximum. At that, with growth of the number of copies of this product, the consumers' wish to pay for it will reduce.

Here comes the contradiction: increase of the network volume leads to increase of usefulness and, this, value of the Internet product, but the price of a single copy reduces. In reality, not everything is as it has seemed at the beginning: total usefulness and value of the Internet product strives to growth, and relative, or marginal, one—reduced during distribution of a large number of its copies. Still, in the period of domination of monopolistic ownership for information data, such dependence takes somewhat different forms. Actually, there is aggravation of the main contradiction of Internet economy's relations between non-commodity content and commodity form of information product.

This process is vividly expressed in the action of the pricing mechanism. According to Sakaiya (1991), price for any product, which technical qualities are unique, will exceed its cost. The researcher notes that there could be no clear connection between prices and basic expenditures for spent materials: manufacturers are able to fix prices at the level that exceeds the cost of the issued products by two-three times, which allows calling their elements the formed knowledge of cost.

But how is it necessary to act within pricing when there's a clear lack of any connection to expenses? T. Sakaiya has a reply for that, as well. According to him, in this case the issue of price formation should be assigned to consumer's idea of "just" creation of price tags. It is necessary to realizes that apart from expenses, there is a specter of other factors that form the buyers' understanding that a certain price is "correct". Of course, an important components of this equation if the price for alternative goods. Apart from all else, social ideas of common sense win. Advertising events and consumers' reviews at various information and communication platforms, expressed by the objects responsible for formation of public opinion, are also important. Sometimes, according to T. Sakaiya, the role of elements of changes is important—as cost, formed by knowledge, is a temporary phenomenon. Thus, T. Sakaiya prefers coming to the following conclusion: from one case to another, price for a specific product might be much higher than the volume of means spent for its production. The difference exists as a result of consumer's ideas' activities.

We think that such reasoning is doubtful as the proofs given within them are based primarily on psychological, not economic, aspects of formation of prices for information products. No one tries to deny the possibility of Internet product's price's being higher that its cost. The true reason of this phenomenon lies in the limited regime of consumption of information products, artificially supported by the most influential IT corporations, and the process of unequal exchange that is based on private monopolistic property for information data.

One of the main peculiarities of the procedure of market pricing in Internet economy is that real pricing takes place not in the production sphere but in the sphere of realization of final products in the market of incomplete competition. Change of the pricing characteristics of Internet product usually faces the artificially created monopolistic limitation and legal acts related to the issues of protection of copyright and intellectual property. Such state of affairs give the owners of information products the chance to receive monopolistic excess profit as a result of realization of Internet products that they control with licensing.

Let us add that pricing for various types of information goods comes from analysis of profitability of the offered information data and market situation. The final result of the process of pricing depends on such factors as expenses for development of product's information content, level of quality of offered data, and expected demand. Apart from that, price for information in entrepreneurship could be calculated on the basis of the following: volume of unextracted profit—due to shortage of commercial data; volume of possible loss from rivals' applying commercial information; volume of profit that company may have in case of possession and further use of commercial data.

It's high time to present the authors' idea of formation of cost and price for information data. It is considered that price should be a reflection of value of a certain copy of information product. Thus, cost and price of the "source" are kept at a high level, which is explained by rarity of intellectual labor expenses used for its manufacture. Cost and price of the second and following copies will decrease proportionally: there's clear logic in that. The larger is the differentiation of unique and demanded information in a specific sphere, the lower is its cost and price that strive to cost and price of the source or time traffic of the Internet use. Thus, the main contradiction of Internet economy is liquidated: with multiplication of Internet product, it gets rid of its commodity form and approaches its non-commodity content. Despite all this, realization of the procedures of establishment of prices in information economy is still far from its ideal.

We have a right to expect that first demand for a new Internet product will exceed its offer. Due to such circumstances, selling the first item of the Internet product for maximal price seems plausible. Later, when the market enters the stage of relative saturation, the price for the Internet product will approach the balance level. However, in real life, price for such products is the average between its high level and the balance level. Obviously, it is a pricing characteristics of the license-protected copy of the Internet product that includes, apart from expenses for copying, money assessment of the copyright and objects of intellectual property of the author.

In practice, manufacturer's monopolization of the issued product, which allows putting a high price, is predetermined by active use of laws on protection of copyrights and objects of intellectual property in this sphere. But even under these conditions, the total profit of the manufacturer would be a share of the possible profit—for the price for a licensed copy of Internet product is lower than its possible maximal price that could seem acceptable even for the richest buyer. Thus, consumer profits are also present in the system.

There could be a situation when pricing in the market of Internet products is different from our expectations—this non-standard situation is created by active involvement of manufacturers of counterfeit products on the Internet. In this case, the initial point is the price consisting of the cost of the material carrier and expenses aimed at distribution of the Internet products' copies. Any price that goes beyond the limits of this level is an obvious profit for so called "pirates".

The viewed circumstances suppose existence of two balance points that symbolize the balance in the market of licensed products and the market of counterfeit products. Not every consumer can but a license product. Obviously, each category of the population has a certain level of expression of buyer's capacity. In view of the above, the demand curve changes. It transforms, turning into a broken line, or divides into several lines, each of which is related to a certain theory of payment capacity. The result is that licensed products suffer from decrease of demand for them—which leads to decrease of the volume of their manufacturer's profit, as at some point its significant share starts to go in the hands of counterfeit's manufacturers. At that, strangely enough, "pirate" activities have a positive side—for the price set in this market is closer to the notion of "justice".

Besides, consumer also receives additional advantages from purchase of unlicensed products: he can have a part of production excess of owners of the licensed copies of information product. Probably, in the near future the leading manufacturers of Internet products will face the problem of effective solution to the problem of holding the positions of their economic domination. Also it may happen that we will witness the decline of current giants of the IT sector—which will be preceded by quick reduction of their capitalization and probable bankruptcy.

2.3 Emergence of Losses and Receipt of Results from Activities of Internet Economy Participants

Over many years, scholars have strived to explain the sense of production and exchange relations, using such categories as production factors—in particular, labor, land, and capital. Once, the great economists Sombart (1930) and Schumpeter

(1939) decided to supplement this range by another position—so called "entrepreneurial capabilities". Still, the dominated position is held by analytical approach to studying the state of economy, within which a special attention is paid to various combinations of labor and capital in the form of labor theory of cost, and the value of knowledge and organizational innovations in management is diminished. However, the conditions under which working hours are reduced and functions of laborers' productions are partially liquidated lead to understanding the fact that knowledge and certain aspects of their practical use can substitute labor as a source of value added. In this respect, labor and capital are among the central economic categories in industrial society, and information and knowledge became the main notions of post-industrial society. Bell (1973) said that when knowledge—in its structural form—becomes involved into practical processing of resources, it is possible to consider that it is knowledge, not labor, is a source of cost.

At present, there could exist other multi-factor models that emphasize the process of increase of the value of technological and information components of production cycles. Some of the still living economists think that scientific and technical progress, which supposes appearance of new types of information technologies, gives the countries with developed economy of industrial type ca. 35% of economic growth—the rest accounts for labor and capital.

Speaking of information resources, it is necessary to note the fact of their quick evolution. Thus, Gilder (1979) was deeply interested in elements constituting the processes of their development—he considered himself to be a representative of the radical technocratic direction of economy and thought that each dozen years the society would face the total reduction of prices for information and communication technologies. Such course of thoughts led the author to the following conclusions: prices for information and technological resources strive to zero.

Indeed, the cost of production of a commodity item of a certain type becomes so low for the subject that sells it that buyers can feel the cost of consumption striving to zero too. Eventually, all of this could be seen in the "Galder curve": price strives to zero, but it can never reach it, for there is a specific minimum price of purchase of a certain information product.

It is clear that manufacturers try not to lose a chance to sell a product for a maximum price. Of course, it is a vivid proof of our statement on the domination of monopolistic discrimination in the market of Internet products.

At present, neo-classic or marginal methodology allows building the graph of marginal cists of information product. At an especially high level, these costs are set during the initial issue of a commodity item. On the whole, the costs related to the product of information character are actually intellectual and labor expenses of creative persons that are expressed in the above products which would be further placed on various types of information carriers or on the Internet.

The costs relate to network benefits differ from the costs peculiar to traditional products. Thus, the main share of costs that account for manufacture of an information product "works" directly at the stage of creation of its very first copy. This means that the gap of disproportion lies between costs of manufacture of the first copy and its following copies. At that, we will deny the chances of reduction of the level of marginal costs as a result of functioning of the saving effect on production scales. The classic theory of economy is based on the law of diminishing returns that substantiates the nature of emergence and existence of many phenomena of economic world. As a matter of fact, information products are not subject to this law. Or, at least, they are peculiar for growing returns in the mid-term and long-term. Thus, sectorial spaces, within the perimeter of which the information products are manufactured, face huge possibilities for participation in the procedures of exploitation of the scale effect.

It is easy to guess that the issue of growing returns was interesting for scholars in the past as well. For example, a well-known representative of neo-classic A. Marshall studied this problem in the spheres where the effect of saving on the production scale could be observed: railway, natural gas production, etc. However, uniqueness of information benefits consists in the fact that profitability grows due to a special structure of costs that account for production of them. Here, the effect of saving on production scales is peculiar for two things:

- unlike the case with the traditional products, in which the work of the production scale effect is gradual and linear, in case of information networks their value grows according to the exponential very quickly;
- then—if the effect of saving on production scales in scenarios with common products is the achievement of self-sufficient organization that could reach it, under the conditions related to Internet products, growth of usefulness is achieved by usual, but multiple, users of the network space. Their multiple character allows the possessing that which we have in this regard.

Recognizing the probability of expenses due to appearance of more perfect variants in the long-term, it is possible to conclude: sooner or later we will witness vivid reduction of prices for goods the process of which requires serious intellectual costs, leading to substantiated reduction of costs as such in the process of prices formation. All of this will lead to decline of the role of offer, formed by marginal costs, and, correspondingly, to actualization of the role of demand.

Stewart (1997) found a lot of differences between contents of cost of knowledge-intensive products and structure of cost of material that received material form. A large share of costs is directly related to the preliminary stage of production—i.e., cost of creation of the first copy of information product and its following copies are connected by disproportionate dependency. So, the lower is the level of product's materiality and the higher is the level of its proximity to real knowledge, the larger is current period's costs' distance from marginal costs. In reality, costs of creation and distribution of electronic copies of a certain documents on the Internet could be compared to an electric spark—though, the cost part in these processes accounts not for the product's recipient but for its manufacturer. Let us remember that manufacturers of industrial products strive to accumulate costs at the initial stage of production with the increase of the volume of their information content. Costs aimed at conduct of R&D and scientific measures during production of various technical items manifest their growth of direct production expenses.

With appearance of the second copy of the information product, marginal costs that take place within its production start reducing to the level of the licensed copy creators, and then—to the level of the counterfeit's copy creators. Of course, there is certain connection here: in the first case, the totality of expenses includes payments to owners of intellectual property. When we speak of counterfeit, we mean bringing down expenses to the cost of cheap information carriers and small costs of distribution of copies on the Internet. At that, the level of initial costs is very low, for creation and further distribution of large volumes of copies require only one information carrier that could be bought by the subject of illegal activities from the legal creator or from a representative of the "illegal world". It is quire probable that the quality of the copy won't be worse that the quality of the original.

Here comes another conclusion: any price, the value of which at least slightly exceeds the volume of marginal costs, poses profit for manufacturers of information counterfeit. The creator of original product will seek the profit in the prices that is higher than the average costs. Despite this, price should be based on just foundations, or, in other words, fully reflect the cost of information product. Otherwise, the circumstances at which the largest IT suppliers are good with growth of prices for products, substantiated by the necessity for protection of copyright, are a driver of aggravation of the primary and secondary contradictions of the network economy. An effective solution in this case could be limitation of realization of licensed copies of information product by either establishing strict time limits or setting certain requirements to quantitative content of distribution of copies, which will allow encouraging intellectual and creative efforts of people that create them. Later, the price for information products will strive to the indicators reflecting the cost of their barriers. Another variant of events: products will be distributed on a free basis, which plays an important role in fighting the phenomenon of intellectual "piracy".

Mutual estrangement of commodity items that are manifestations of average labor costs is replaced by the phenomenon of distribution of costs. The larger is the circle of persons interested in consumption of information resources for the purpose of desobjectivation, the larger is the total volume of unit costs used for their production.

Thus, the price after the end of the relevant period should transfer from the level of the price for a licensed copy to the level of material carrier copy or Internet traffic, and then "bring itself to zero"—i.e., stop at a zero level. The resulting profit of licensed products' owners will become a good bonus base for their developers and manufacturers.

Again, one of the main contradictions of the network economy, strangely enough, is manifested in institutional traps. For example, one of these is popularization via the Internet of software, effective functioning of which requires corresponding compatibility with other programs. At that, initial distribution of information products via the global information and communication network could be absolutely free—however, in the future periods the consumers will have to pay for the consumers IT products—and pay a lot. In case of strict observation of the rule of "just" pricing in the "new" economy, such trap could be easily avoided—except for the relevant period.

Based on the above, it is expedient to pass to conclusions on this part of the work.

Conduct of various events for studying the processes of development of Internet economy at micro-level supposes their organizers' realizing the fact of presence of significant differences between Internet service and Internet product. Thus, the former, being a usual commodity, is limited by time and space limits. Internet product does not always have to have a commodity content. Its consumer cost consists in its being applied as a means for creation of other commodities of material and non-material world.

Value and cost are the main characteristics that present Internet product as an item of the system of monetary exchange that continues existing in our days due to presence of a wide sector of material production and using in Internet economy the intellectual labor—very rare and useful resource. At that, traditional theories of economy—labor theory of cost and theory of marginal usefulness in their original form—are not always good variants for substantiating the sense of product exchange processes in Internet economy. Moreover, other theories of cost (energy, information) are not good either.

Product of information content acquires commodity form when the issue of its realization directly depends on the quality of artificial limitation of access to the data constituting the integrity of such, with various tools of online protection (PIN, licenses, etc.). All of this provokes just growth of contradiction between non-commodity content and commodity form of Internet product, based on appearance of the transformed form of cost of all copies following the first copy of information product.

Thus, relations of Internet economy is a system of quickly developing interactions, which is something larger than a phenomenon that goes beyond the power of the main economic laws: cost, marginal usefulness, demand and offer, and diminishing feedback of production factors.

Chapter 3
Perspectives of Internet Economy Creation

3.1 Advantages of Internet Economy

Internet economy offers deep changes of the foundations of socio-economic systems. It gives economic processes and relations completely new qualities and characteristics: from organization of production to marketing and sales, and from creation of business to making purchases. Peculiarities of Internet economy give it multiple advantages, which are vividly seen against the background of classic economy (Table 3.1).

Let us view the advantages of Internet economy in detail. Internet economy uses the Just in time (JIT) approach to organization of production and distributive processes. It supposes starting the production process right after the customer's making an order.

Due to this, the necessity for storing resources and final products is eliminated, and manufacture of the required volume of products is ensures—that is, the balance of demand and offer in the market is achieved.

In comparison, classic economy uses the Just in case (JIC) approach. Within this approach, the company manufactures products, either depending on the lower limits of production capacities (including striving to getting the scale effect) or on the forecasts for sales of final products on the basis of marketing studies.

After that, the company tries to sell the manufactured products in the market, which is not always possible, so the company is obliged to offer discounts (which can damage business), or utilize or process unsold products.

In their turn, consumers cannot place individualized orders, i.e., ask to given the purchased products specific characteristics, and have to select the already manufactured products, which qualities cannot be changed. This does not allow satisfying existing needs of the society in full.

Internet economy optimizes the process of interaction between sellers and buyers in the market. Internet companies can offer a very wide assortment of products, which expands the possibilities of choice for consumers. After collecting the orders,

A.P. Sukhodolov et al., *Internet Economy vs Classic Economy: Struggle of Contradictions*, Studies in Computational Intelligence 714,
DOI 10.1007/978-3-319-60273-8_3

Table 3.1 Substantiation of advantages of Internet economy in contrast to classic economy

Viewed aspects of economic processes and relations	Type of economy		Advantages of Internet economy
	Classic economy	Internet economy	
Organization of production and distributive processes	Approach just in case	Approach just in time	Establishment of market balance
Barriers of entering the market	High institutional limitations	Low institutional limitations	Higher level of competition
Dominating form of business in economy	Traditional business	Internet business	Lower expenses of business
Coverage of business	Geographical boundaries set limits on business	Global coverage of business (not limited by geography)	Wider possibilities for sales
Time limitations of work of enterprises	Strictly set regime of companies' work	Companies work 24/7	Expanded possibilities for service
Complexity of consumers' making decisions	Consumers do not know all offers in the market	Consumers have full information of offers	Optimality of consumer choice
Dominating production and sales (marketing) strategy	Specialization (general approach for all customers)	Diversification (individual approach to each customer)	Full satisfaction of society's needs
Dominating form of performing payments in economy	Financial operations with cash transactions	Non-cash transactions	Elimination of shadow economy

Internet companies give them to manufacturers, which, in their turn, can cooperate with several distributors—accepting orders from them they can use their production capacities in full.

Using the JIT approach to organization of production and distributive processes will allow reducing the risk components of entrepreneurial activity. As it is one of the key restraining factors on the path of starting business, Internet economy stimulates development of entrepreneurship. This also allows reducing the product's price, as expenses for risk management are shifted to consumers—i.e., they are included into the price.

Another advantage of Internet economy is low barriers for entering the market. Here we speak of institutional limitations. While in case of creation of traditional companies, the state's support is required in a lot of spheres of economy (permission for participation of private business in provision of this type of goods or services) and it is necessary to get licenses or go through inspections (fire safety requirements, etc.), Internet business is free from that.

Anyone can create an Internet site and commercialize it (place an offer of goods or services). This allows eliminating the influence of state institutes which are usually ineffective due to a wide bureaucratic machine and corruption component of entrepreneurial activity regulation.

Elimination of these institutional imitations simplifies the procedure of creation and doing business and reduces its expenditures, as well as the price for the products offered for consumers, as additional expenses of enterprises are shifted to them. Thus, Internet economy stimulates the development of the service sphere, especially trade.

Due to reduction of market barriers, Internet economy achieves higher level of competition. This stimulates high effectiveness of entrepreneurship which strived to reduction of expenditures of business and provision of attractiveness of commercial offers with the help of low prices, unique offers for customers, active marketing activity, etc. That is, business constantly develops—otherwise, it might go bankrupt.

The dominating form of business in Internet economy is Internet business. Creation and successful work of an Internet company require a minimal set of resources. It is necessary to officially register the company, develop and start the Internet site, and conclude an agreement with products' suppliers and shipment organizations (develop shipment of products to customers).

It is possible to conduct business at home, without any hired help. Of course, this is the simplest scheme of starting an Internet company—which can be expanded. It is possible to conclude an agreement with a financial organization (as a rule, a bank) for providing the possibility of paying for the products through its web-site (Internet acquiring).

Classic economy uses a much more complex scheme. In order to start a traditional company, it is necessary to register it, rent (or purchase) premises for the company, buy (or produce) products, attract consumers, hire personnel, etc. This requires significant resources: time, material, financial, and human.

The minimal necessary set of resources for creation of a company in Internet economy allows involving a large number of people into business. This stimulates provision of self-employment of population and creation of additional jobs, which leads to reduction of unemployment rate.

In classic economy, geographical limits set limitation on business. Binding to premises, equipment, and human resources determines low mobility of the traditional entrepreneurship. This predetermines its high dependence on the situation in territorial markets in which it is present.

Internet economy ensures the global coverage of business (not limited by geography). An Internet company can offer its products all over the world—for everybody has access to the Internet. Higher capacity of the world markets allows a company to get advantages from the "scale effect".

If the interest to its products on one territory reduces, the company may shift (redirect its marketing efforts) to other territories. Apart from higher flexibility and mobility of business, Internet economy ensures larger possibilities for optimization of production and distributive processes in entrepreneurship.

While in classic economy, only certain entrepreneurial structures can transnationalize (distribute their production and sales networks in various countries of the world), every company in Internet economy has such possibility. Moreover, it is not necessary to create branches and departments in order to be present in different countries—it suffices to conclude agreements with local business structures.

This allows for development of products on the country of residence with highly qualified human resources, production in the countries with good access to material resources and low cost of human resources, and sales in any place on the planet.

Internet economy allows eliminating time limitations of companies' world. In classic economy, the working hours of companies are strictly set—most of them work during the week-days (Monday–Friday). Most of customers also work on these hours, so they cannot obtain full information on existing offer in the market or make the optimal choice. This violates the condition of rationality of economic operations.

Internet companies are open 24/7. Due to this, consumers can learn their assortment when they wish. Moreover, they can make an order and even pay it (with the help of Internet acquiring system). The customers do not have to spend their time, effort, and money for visiting stores—they can select products and make purchases while at home.

Additional working hours allow business to successfully sell products all over the world, which is impossible in case of traditional enterprises due to time difference. This, an Internet company can create only one Internet site, selling products in different countries. A traditional company would have to open a lot of branches in each country.

Maximal comfort of purchases provides expanded possibilities for service in the conditions of Internet economy. At that, this does not require additional expenses from business, but saves its resources, eliminating the necessity for additional personnel, premise, equipment, etc.

Attractive service creates favorable conditions for a large number of purchases, stimulating sales. More deals are concluded in Internet economy than in classic economy. This stimulates the development of entrepreneurship and accelerates economic growth (increases the rate of GDP growth).

Difficulty of decisions making by consumers in Internet economy is reduced. They have full information on the offer due to its availability on the Internet and presence of effective schemes of search and sorting of electronic information. These systems allow determining suppliers of the sought good or service and comparing the offers according to the set criteria (prices, technical and marketing characteristics).

In classic economy, consumers do now know of all offers in the market and often make purchases according to the geographical criterion (close to home or current location). Due to this, a winner of competitive struggle is the one with the better location—which is an important, but not the main, indicator of products and services' quality.

Internet equals conditions for all market players. That's why, instead of struggle for better location, entrepreneurs compete for quality of Internet sites, quality and price for products, methods of its shipment and payment. Under the conditions of Internet economy, consumers always make the optimal choice or have such a possibility, while in classic economy the condition of optimality is violated due to incomplete information and complexity of its processing, called the "limitation of calculation capabilities of a human".

It is worth noting that representatives of the classic economic school who confirm the necessity for rationalization of choice and offer a model of market economy for achieving this goal, state the impossibility of its practical achievement. Internet economy allows for eliminating this contradiction and for realization of the most perspective provisions of the classic school of economy.

In Internet economy, the dominating production and sales (marketing) strategy is diversification which supposes an individual approach to each customer. That is, it is not the enterprise that determines the conditions of purchase but consumers—and each deal is unique. It is important to note that strong power of consumers in the market does not created impossible conditions for the companies that lead to them to bankruptcy but stimulates healthy competition of business structures, as they are adapted to such high flexibility.

Traditional companies use the strategy of specialization that supposes a common approach to all customers. On the one hand, it simplifies a lot of business processes due to wide possibilities in the sphere of their standardization. On the other hand, it does not allow them to fully correspond to the requirements of each client and to fully satisfy his needs. At that, in classic economy companies are not flexible in the short-term and the long-term, which hinders the development of business and economy on the whole.

Internet economy has all conditions for constant innovational development of entrepreneurial structures. This stimulates formation of innovations-oriented economic system, competitiveness of which in the modern world is much higher that with the system that preserves tradition. Thus, Internet economy stimulates optimization of internal economic processes and strengthening of the country's positions in the world arena.

Internet companies are able to establish a close contact with consumers for managing their loyalty, which gives them wide marketing capabilities. In Internet economy, companies can collect a lot of information on their customers, which allows them to optimize assortment of products. They can also inform their customers on their new products and special offers with minimum efforts.

A dominating form of transactions in economy—non-cash financial operations. This ensures their transparency and control by the state. Non-cash transactions are easily tracked, and the information on their participants is easily obtained. Classic economy is dominated by financial operations with cash transactions, which cannot be tracked—which creates preconditions for development of shadow economy. In Internet economy, the very possibility of shadow deals is absent.

Internet economy also stimulates the increase of corporate social responsibility. The payment for labor of companies' employees is performed with the help of non-cash transactions. This ensures open character of populations' income and full payment of social payments by employers, including contributions into the social insurance fund.

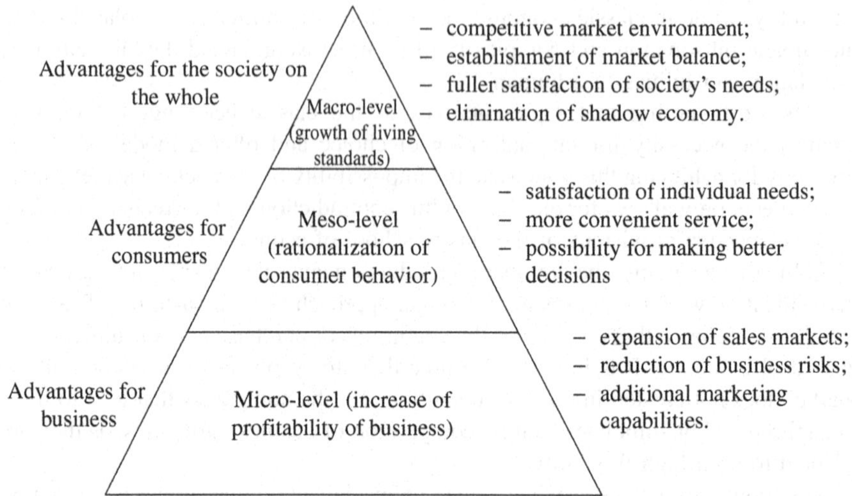

Fig. 3.1 Profit of various beneficiaries from advantages of Internet economy at different level of the economic system

The possibilities for corruption also disappear, as all performed payments are reflected in information systems of their providers (as a rule, banks). At that, the state's expenses for control of the financial system of the country and for fighting the shadow economy are reduced. That is, the budget assets are saved—and these assets could be spent for other important projects.

Due to that, the state budget receives tax payments from companies and individuals. This expands financial possibilities of the government and creates additional perspectives for conduct of socially important programs and projects. In the long-term, this stimulates increase of the living standards of the population. The whole society can feel the advantages of Internet economy.

The determined advantages of Internet economy provide the corresponding profits for various beneficiaries at various levels of the economic system (Fig. 3.1).

As is seen from Fig. 3.1, the beneficiaries of Internet economy's advantages at the micro-level are business structures, for which the growth of profitability of business is ensured. At the meso-level, the beneficiaries are consumers who can make a rational (optimal) choice. At the macro-level, increase of effectiveness of economy and living standards is ensured.

Based on the above, it is possible to conclude that Internet economy improves (simplifies, accelerates, and increases effectiveness) of business processes of purchases and stimulates achievement of various socio-economic goals. That's why Internet economy is not only a preferable form of organization of economic activity but also a means of solving the actual problems of society and economy.

3.2 Threats and Problems of Creation of Internet Economy

The determined advantages of Internet economy substantiate the expedience of its creation. However, as any other socio-economic process, it has to be related to various problems and threats. In order to have a comprehensive idea of Internet economy, let us view its potential drawbacks and evaluate the perspectives of their elimination.

The most important problem of Internet economy is security. More than 30% of crimes in the world economy are performed in cyber-space. As is expected, their quantity in the future will be growing (Marcus 2011). According to the official data of the Main information and analytical center of the Ministry of Internal Affairs of the RF, the number of crimes in the sphere of computer information constituted 1739 in 2014, reaching 2382 in 2015 (Ministry of Internal Affairs of the RF … 2017). Dynamics of the volume of internal Russian market of Internet business in 2011–2016 is given in Fig. 3.2.

As is seen from Fig. 3.2, in 2011–2016 the volume of internal market of Internet entrepreneurship in Russia was growing by 27% annually. This proves shows gradual transition of business into Internet space and creates preconditions for growth of cybercrimes. Thus, a threat to information security and financial security emerges—at that, they both influence consumers, entrepreneurs, and society on the whole.

A threat to information security is caused by the electronic data's aptitude to attacks by cyber criminals. In classic economy, anonymity of deals is provided by lack of collection of information on their participants (except for the concluded

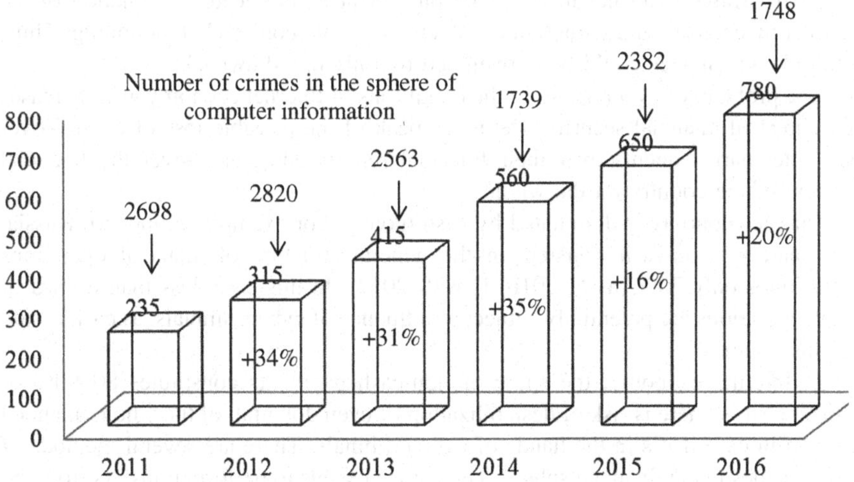

Fig. 3.2 Volume of the internal Russian market of Internet entrepreneurship in 2011–2016, RUB billion. Compiled by the author on the basis of the data of C-News (2017)

contracts). That is, when a buyer comes to a store, he can buy a product without sharing any information about himself and paying cash for the product.

In Internet economy, each deal leaves a trail that contains detailed information on the participants and the deal. This information is confidential and is protected by law and security systems. However, there's always a possibility of disclosure of this data. Surely, not all the data pose value (the possibility of getting profit from possessing them) and interest for cybercriminals.

For example, there's not profit from the information that an unknown person bought groceries in a store yesterday. At the same time, the information that a large company that employs half the town took a multimillion credit from the bank, accumulated a large debt, and is at the edge of bankruptcy may lead to a catastrophe in the labor market and investment market, changing the price for stock of this company—which would be profitable for cybercriminals.

In many cases, a formal or informal condition of conclusion of an agreement is preservation of confidentiality—if its participants are individuals, or commercial secret—if its participants are legal entities. That's why the possibility for disclosure of the information and its further use against the owners can be a reason for refusal from the conclusion of the agreement.

In this context, creation of Internet economy can be a restraining factor of development of economy, being a cause for reduction of a number of deals—in spite of increase of their commercial attractiveness and convenience. In the scale of the state, disclosure of information that poses importance and value in the national scale, may lead to the economic crisis.

Apart from stealing the information, cybercriminals can issue disinformation, providing false data regarding certain events, subjects, or objects, thus misinforming the market participants. Cybercriminals can also hide important information regarding themselves or others.

Apart from certain negative social consequences, this leads to violation of the market mechanism and reduction of effectiveness of economy's functioning. Thus, multiple advantages could be outweighed by only one drawback.

The problem of cybercrimes in the conditions of Internet economy also increases the threat of financial security. Here we speak of the possible loss of control over separate market agents over their financial reports and state—over the financial system of the country on the whole.

Classic economy is dominated by cash money. For example, in modern Russia, the share of non-cash transfers in the general structure of financial operations constitutes only 30% (as of 2016) (TASS 2017). In this case, less than a third of such operations are potentially subject to influence of cybercriminals, as cash is safe in this regard.

In Internet economy, the share of non-cash payments constitutes 80–90% or more. That is, there is risk of destabilization or even downfall of the whole financial system in case if it's in the hands of cybercriminals. There are several methods of cybercrimes in the financial sphere. Thus, it is possible to perform transfers from the accounts of the organization and individuals to the accounts of the cybercriminals.

That is, here we speak of stealing money from the owners. Apart from enrichment of cybercriminals, there's another side related to this process—decrease of population and companies' income that leads to decrease of living standards and growth of financial risks for business.

Using the emission mechanism is also possible, which leads to growth of the volume of money in turnover. Thus, cybercriminals do not steal financial assets but increase the volume of their own assets on their electronic financial accounts. An obvious result of this is growth of inflation in the country, which damages population and entrepreneurial structures.

Another means of financial cybercrimes is stealing money from the state. In this case, it won't be able to perform its function, and the whole country will face external threats and internal problems related to termination of social programs' realization.

The methods of financial cybercrimes also include disruption of the world of financial organizations (primarily, banks). Apart from damage to these very organizations (owners of business, employees, and intermediaries), this includes reduction of opportunities for development of economy, bereaving it of investment and credit resources.

Another problem of creation of Internet economy is the fact that not all products could be selected on the Internet. For example, household appliances can be selected on the basis of specified characteristics, while the most detailed information on clothes is not enough for seeing it fit a specific person.

The very specifics of many products make their selection via the Internet uncomfortable. Therefore, full elimination of traditional enterprises and creation of a 100% Internet economy is impossible. Thus, a threat to polystructure of economic system arises within which it is dominated by Internet companies but preserves a large share of traditional companies.

This does not allow getting full advantages of Internet economy and complicates the process of management and regulation of socio-economic processes in it. Depending on consumer preferences, demand for the traditional companies might preserve or even grow. This puts in doubt the very possibility of creation of Internet economy in practice.

A partial solution of this problem is creation of a network of demonstration centers in the whole country or territory on which the company orients. Thus, consumers can see the existing assortment of products or select the required product in these demonstration centers, and then buy it from the Internet company with a possibility of receiving it in this center.

This allows bringing down the quantity of traditional companies to the minimum; in view of the fact that Internet economy has the market structure of competition, the market has a lot of small and medium business structures, and of each of them creates its demonstration centers, the share of the traditional entrepreneurship will be high.

The problems of creation of Internet economy include also the delayed character of product shipment from the moment of order and payment. In many cases, consumers are ready to wait for the product as long as it takes, but there are situation, when the need for the product is urgent. In case with medicine, long shipment terms might lead to severe consequences for the buyer's health.

This also sets limitations on the possibilities of creation of Internet economy and predetermines the necessity for preservation of at least a part of traditional enterprises. A partial solution of this problem is development of a system of shipment of Internet products. This will enable urgent shipment—1-day shipment.

Though the very concept of JIT, which lies in the basis of work of Internet companies, supposes not only the delayed shipment but also the delayed manufacture of products. For the purpose of optimization of production, manufacturers will be collecting orders over a long period of time, and will start the production only after they have the necessary number of orders.

This leads to necessity for preservation of the JIC system in the conditions of Internet economy creation. As a matter of fact, in order to build Internet economy in one country, it is necessary to preserve classic economy in another. In the scale of meso-economy, some regions can build Internet economy on the basis of classic economy in other regions as a part of the national economic system.

This sets serious limitations on the potential scale of Internet economy and makes its formation in the global scale impossible. Therefore, the advantages from its creation will be accessible not for all participants of the world economic system.

One of the problems of building the Internet economy is difficulty with a product return. While one can easily come back to a traditional store and get the money back or replace the product, in order to return the product purchased from the Internet company a consumer has to use the shipment system, which leads to additional expenses.

Besides, the consumer will have to wait for return of the paid money, as the company won't return it until it receives the product back. Accordingly, the consumers may be interested in potential advantages of Internet economy, but when facing its drawbacks they might be disappointed in its very concept and demand the return to classic economy.

The problems of building Internet economy also include instability of business. In the conditions of Internet economy, business risk increase. Firstly, the risk of provision of information and financial security grows. This requires from entrepreneurial structures additional expenses for managing these risks.

Secondly, the reputational risk of business structures grows. It is caused by the fact that in the conditions of Internet economy the speed of information transfer is very high, and its accessibility for wide masses of population is large. Thus, while in the conditions of classic economy the companies can show their advantages and hide disadvantages, in the conditions of Internet economy it is much more difficult—for consumers and other interested parties learn everything that is related to the company's activity.

Thirdly, the risk of changing the market situation grows. The consumers who possess high power in the market in the conditions of high competition among sellers, constantly change their preferences, which predetermines high volatility of demand. Even flexible small and medium Internet companies need to adapt depending on changing consumers' preferences.

According to this, a threat of instability of entrepreneurship arises—in particular, of constant bankruptcy of some and creation of other Internet companies. Lack of orientation at the long-term period of their existence can become a reason of irresponsible approach to doing business.

This might be expressed in irresponsible attitude towards hired help—unpaid wages, non-provision of social guarantees, etc., as well as irresponsible attitude towards consumers—untimely shipment or failure of shipment of paid products, low quality of products, etc.

It should be noted that such measures do not contradict the effective law due to impossibility of lodging claims to bankrupt companies. That is, the state cannot protect consumers and employees whose interests were damaged by such irresponsible actions of Internet companies.

Another problem of creation of Internet economy is its instability. Domination of the service sphere—in particular, trade—makes Internet economy dependent on Internet companies which develop quickly but function in the conditions of high global competition. Without a strong real sector, economy will be unstable and subject to crises.

In the classic industrial or even post-industrial economy, a large share of industry is preserved. This allows for formation and support for competitive advantages of economy in the long-term, as demand for industrial products is characterized by sufficient stability.

In Internet economy, the service sphere ousts all other spheres from the country's GDP. It is more subject to economic crises due to instability of demand for services and domination of small and medium entrepreneurship in the service sphere. Thus, in case of economic crisis, the service sphere is more subject to its negative influences.

The determined problems of creation of Internet economy and related threats are shown as a system in Fig. 3.3.

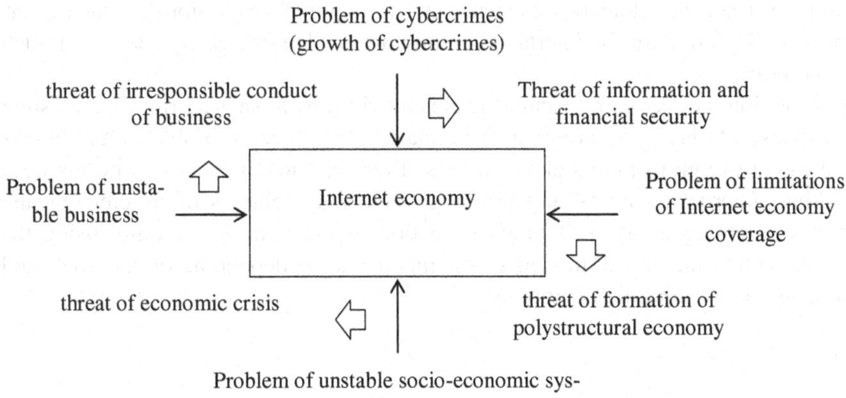

Fig. 3.3 Threats and problems of building Internet economy

Table 3.2 Results of SWOT analysis of Internet economy creation

Advantages of Internet economy	*Drawbacks of Internet economy*
– Growth of living standards – Rationalization of consumer behavior – Increase of profitability of business	– Problem of security – Problem of limitation of coverage – Problem of instability – Problem of unstable business
Perspectives of Internet economy	*Threats to Internet economy*
– Development of technologies for provision of Internet security – Establishment of the balance of Internet entrepreneurship and traditional entrepreneurship in Internet economy (their simultaneous existence in the required proportions)	– Threat to information and financial security – Threat to irresponsible conduct of entrepreneurship – Threat to formation of polystructural economy – Threat of emergence of economic crisis

As is seen from Fig. 3.3, each problem related to Internet economy causes a corresponding threat. Based on the viewed advantages and drawbacks of Internet economy, it is possible to have its comprehensive picture and to determine perspectives of elimination of these threats, i.e., conduct SWOT-analysis, results of which are given in Table 3.2.

As is seen from Table 3.2, perspectives of Internet economy are related to creation and implementation of innovational technologies that allow for provision of the high level of protection of the information and financial system from attacks of cybercriminals. In view of constant renovation of attacks' methods, development of technologies for supporting information and financial security should be continuous.

Practical impossibility of full elimination of traditional entrepreneurship in the conditions of building Internet economy leads to the necessity for establishment of the balance of Internet entrepreneurship and traditional entrepreneurship in Internet economy (their simultaneous existence in the required proportions). That is, perspectives of formation of Internet economy are predetermined by search for such proportions.

Thus, Internet economy cannot be characterized as a panacea from the existing drawbacks of classic economy, as it is related to its specific problems and threats that can outweigh its potential advantages. Therefore, maximization of profits from building Internet economy and bringing down the problems of its creation and threats emerging as a result of its formation to the minimum require using the corresponding theoretical models of Internet economy depending on the goals and existing socio-economic conditions.

3.2.1 Theoretical Models of Internet Economy

In practice, there could be a lot of variants of combination of the traditional and Internet entrepreneurship within creation of Internet economy. Generalizing these variants, it is possible to distinguish the main theoretical models of Internet economy, comparative analysis of which is performed in Table 3.3.

Let us view the distinguished theoretical models of Internet economy in detail. The pure model supposes full absence of the traditional entrepreneurship. This is ensured by means of close cooperation with other countries of the world. That is, service sphere—primarily, trade—is developed in a country, and manufacture of sold and consumed products are performed in other countries.

From the point of view of the global economy, realization of such model of Internet economy in some countries and preservation of classic economy in other countries is rather perspective and fully corresponds to the interests of getting profit for all participants of the global economic processes from the international division of labor.

Table 3.3 Comparative analysis of theoretical models of Internet economy

Criteria of comparison	Model of Internet economy		
	Pure	Dominating	Doubling
Role and importance of Internet entrepreneurship	Lack of the traditional entrepreneurship	Domination of the Internet entrepreneurship	Doubling of the traditional entrepreneurship
Spheres of economy presented in the structure of GDP	Only service sphere	Service sphere on certain territories, and other spheres on other territories	All spheres of economy
Organization of production and distributive processes	Only the JIT approach	Both the JIT and the JIC approaches	
Independence of economy	Low	средняя	High
Sustainability of economic growth and development	Low	средняя	High
Targeted countries	Developed countries with post-industrial economy (USA, France, Spain, etc.)	Developed and developing countries in economies of which service sphere and industry have an important role (Japan, Germany, etc.)	Developing countries with industrial specialization of economy (Russia, India, China, etc.)

That is, all countries with high effective demand can develop Internet economy, and the countries with cheap human and excessive material resources can develop production and perform orders. While considering only the service sphere and industry, it is possible to observe that such structure of the global economy works at present.

Such clear division of roles of the countries in the global economy and strict division of labor are related to drawbacks related to the industrial countries' dissatisfaction with their low position and their unwillingness to solve socio-ecological problems that are consequences of active industrial production.

From the point of view of the countries that realize the model of the pure Internet economy, it possesses both positive and negative features. On the one hand, efficiency and effectiveness of economy are maximized. On the other hand, its strong dependence on external suppliers of the finished products is established due to full elimination of internal industrial production. That is, it is possible to see division into consuming and manufacturing countries in the global economy.

In case of violation of international economic relations due to the global crisis and violation of the normal system of connection through which the information is transferred to the countries that realize the model of classic economy and perform orders from the countries with Internet economy, or sudden delays in shipment of finished products, an urgent deficit may appear. This predetermines high risk of systemic crisis in the model of pure Internet economy.

Using only the JIT approach during organization of production and distributive processes creates additional difficulties for consumers, as they have to wait for the place and paid orders to be transferred to countries-manufacturers, for they have to collect the optimal number of orders, manufacture and transport the goods.

This shows doubts in the true power of consumers in the theoretical model of the pure Internet economy, as the traditional companies in classic economy optimize their activities and get larger profit. Therefore, it is possible to state that there are substantial positive externalities related to creation of Internet economy in some countries in the world and preservation of classic economy in other countries.

However, while addressing the problem of differentiation of the level of socio-economic development in countries of the world and the problem of disproportions of economic growth in the modern global economy, the theoretical model of the pure Internet economy does not stimulate their solution, but aggravates these problems and increases contradiction of interests of the modern global economic system's participants.

This model fits the developed countries with post-industrial economy (USA, France, Spain, etc.), which perform practical measures for eliminating the industrial production from their territory and specialization in the service sphere.

Theoretical model of the pure Internet economy is shown in Fig. 3.4.

As is seen from Fig. 3.4, Internet economies have a dominating position—they dictate classic economy what to produce. In Internet economy, customers set their specific needs. Internet companies collect all orders and transfer them to classic economy, where the orders are sorted and transferred to the traditional companies.

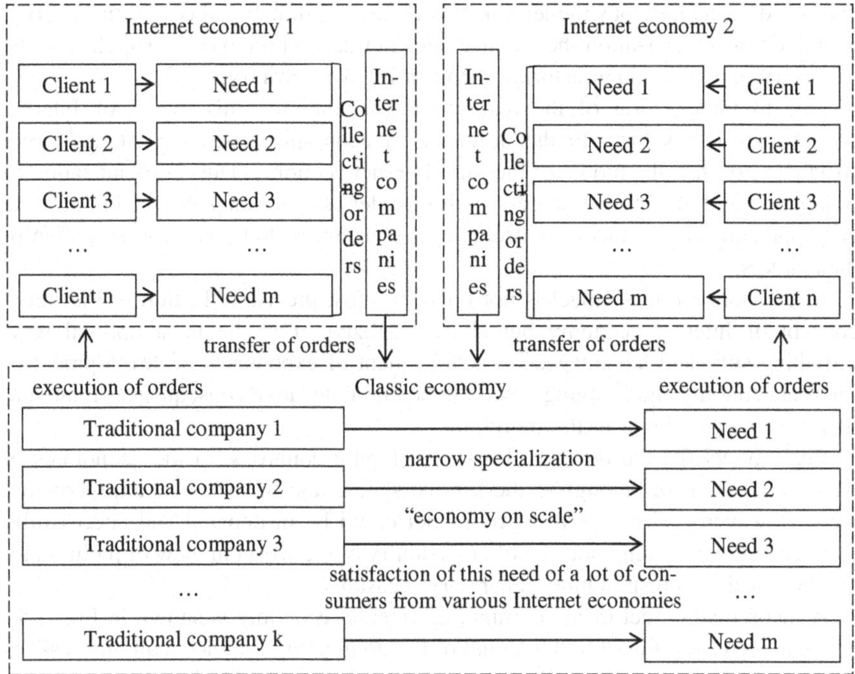

Fig. 3.4 Model of the pure Internet economy

Each traditional company is characterized by the narrow specialization at satisfaction of a certain need of customers. Having collected the required number of orders, they start the production process and manufacture the products, "saving on the scale", satisfying the need of multiple consumers from various Internet economies. After that, the finished products are transferred to the customers through Internet companies, which in this scheme are intermediaries in the system of interaction of manufacturers and consumers.

The theoretical model of the dominating Internet economy received its tile due to the fact that it ensures domination of Internet entrepreneurship with preservation of a large share of the traditional entrepreneurship. Thus, certain territories are dominated by the service sphere, while other develop other spheres of economy.

Obviously, the disproportions in development of the global economic system that were discussed in the context of the model of pure Internet economy go to the national scale in the model of the dominating Internet economy, preserving their sense and urgent character. Undoubtedly, the territories that are obliged to give away their interests of modernization and take ecological and social expenses of industrial production will be not satisfied with their position.

That's why the theoretical model of the dominating Internet economy is potentially related to the growth of internal contradictions of the national socio-economic system and the risk of its destruction. At the same time, as compared to the previous model, this one ensures high national economic security, as in case of its realization the country does not depend on external supplies of the necessary products, manufacturing them on its own territory.

Due to preservation of the real sector of economy, this model of Internet economy is less subject to the influence of economic crises, i.e., it has larger sustainability than the model of the pure Internet economy. During organization of production and distributive processes, this model uses not only the JIT but also the JIC approach. This reduces the problem of consumers' long wait for execution of their orders.

In our opinion, this model is not only the most precise reflection of the initial concept of Internet economy but allows realizing it in the most optimal way, providing sufficient advantages for achievement of expedience of development of Internet economy and bringing down the potential negative consequences related to its problems and threat to the minimum.

Such model fits the developed and developing countries, in the economies of which a large share belongs to the service sphere and industry (Japan, Germany, etc.). However, we're sure that this model could be in demand and successfully realized in the economic practice of any country of the world in view of preliminary conduct of the corresponding preparatory measures.

A theoretical model of the dominating Internet economy is shown in Fig. 3.5.

As is seen from Fig. 3.5, the model of the dominating Internet economy realizes the same production and distributive scheme as in the pure Internet economy—not in the international but in the national scale. That is, separate countries are replaced by different territories in one country.

In this case, it is possible to speak of creation of Internet economy only on separate territories, while classic economy will remain on other territories. However, Internet economy is established in the national scale.

The doubling Internet economy differs significantly from the above models and supposes another approach to organization of production and distributive processes. In it, the Internet entrepreneurship doubles the traditional entrepreneurship—i.e., it exists together with it and is to ensure achievement of the goals of its strategic development.

The process of creation supposes creation of Internet companies on the basis of existing traditional enterprises with preservation of the latter in the unchanged form. Analogs of this scheme could be seen in the economic practices of the countries of the world. Thus, rather large companies with necessary resources, long-term orientation at presence in the market, and with to strengthen their reputation create Internet sites on which consumers and interested parties can learn the full assortment and description of their products and purchase them.

This tendency is a reflection of transfer of the competition between production and distributive companies to a new marketing level—they compete not in the aspect of price or quality of products but in the aspect of methods of their promotion and sales. Own web-site becomes an inseparable component of companies' competitiveness in many sectorial markets.

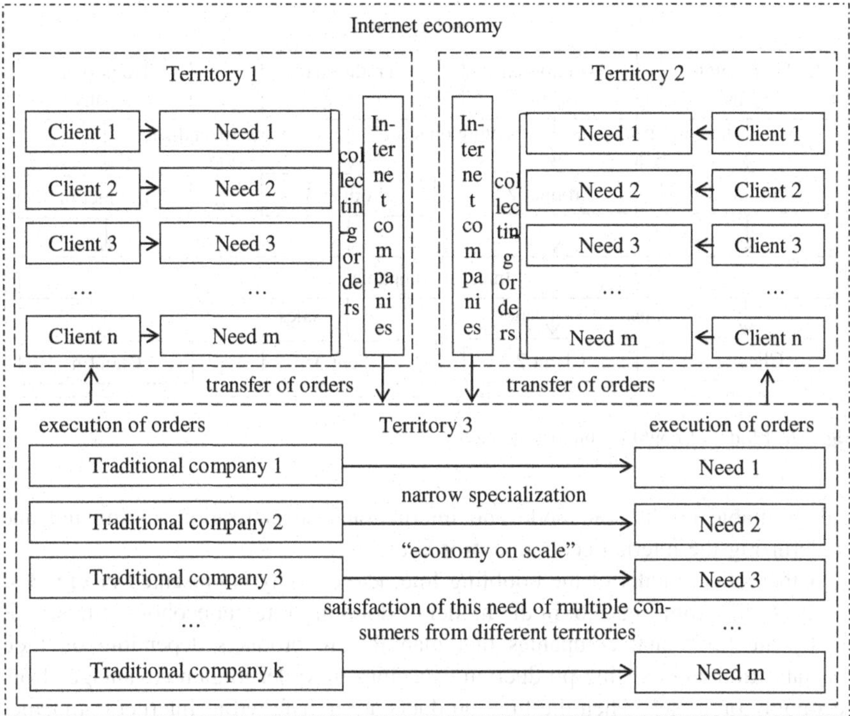

Fig. 3.5 Model of dominating Internet economy

That is, during realization of the model of doubling Internet economy all spheres of the economic system are preserved, and the service sphere goes to a new level, performing not only the function of provision of services but also promotion and sales of products of participants of the previous stages of production and distributive process.

Due to full preservation of the structure of economy with small modification, maximal sustainability of economic system is ensured, as the real sector is not just present in the economy, but occupies a substantial share. Also, the highest possible independence of the economic system from the external world is achieved—as far as it is possible in the conditions of globalization and integration of the modern global economy.

Despite the fact that this model of Internet economy uses the JIT and the JIC approaches at the same time, the latter prevails. This guarantees not only timely but even preliminary execution of most orders of consumers before their placement. At the same time, the possibilities of considering the individual needs of the customers are reduced.

The targeted countries for building a theoretical model of the doubling Internet economy are developing countries with the industrial specialization of economy (Russia, India, China, etc.). Such model will allow them to preserve the traditional

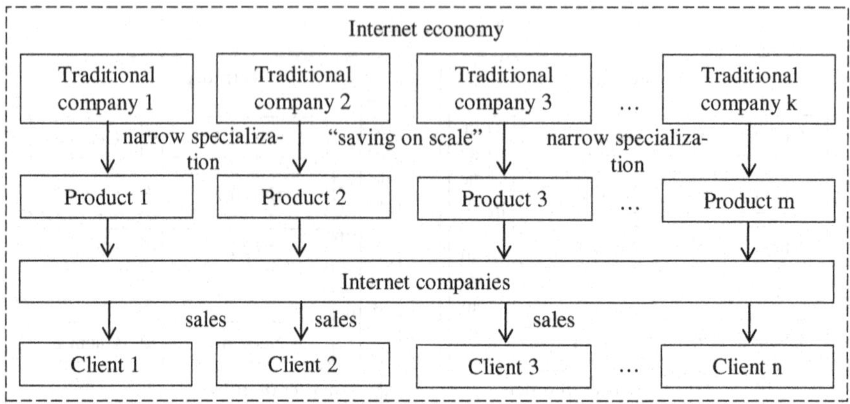

Fig. 3.6 Model of doubling Internet economy

and profitable production mode and international specialization, optimizing and modernizing the internal economic processes.

A theoretical model of the doubling Internet economy is presented in Fig. 3.4.

As is seen from Fig. 3.6, in the model of doubling Internet economy the tone is set by the traditional companies that manufacture products depending on their specialization and existing production capacities, in order to gain advantages from "economy on scale". Then the final products are transferred to Internet companies which ensure sales of the products to consumers.

In this production and distributive system, individual needs of customers go to the background and are not always taken into account by manufacturers of the products. Therefore, the power of manufacturers is much higher than the market power of consumers. Internet companies perform sales function in this chain, not collecting orders from the consumers but accepting products from the manufacturers.

This scheme is close to classic economy, in which Internet companies began to appear. Of course, there are practical examples of realization of this scheme—for example, in modern Russia. It is harmonically built into the existing industrial economy.

It is possible to conclude that there are no significant limitations on creation of Internet economy depending on the production specialization on the stage of development of the national economy. Each modern economic system can select the required theoretical model of Internet economy depending on existing socio-economic conditions and advantages that it's going to achieve and drawbacks that it is ready to accept or eliminate.

Based on the general results of the research, it is possible to conclude that creation of Internet economy is a complex process. Inattentive approach to it may lead to severe consequences for the socio-economic system. At the same time, with a well-planned and effectively managed approach, creation of Internet economy can provide the economic systems with multiple internal profits related to increase of living standards and development of entrepreneurship, as well as the desired role in the system of international division of labor.

The main obstacle on the path of creation of Internet economy in modern economic systems is its vulnerability to fluctuations in the world markets and global crises, as well as imperfection of existing technologies of provision of data and financial operations' security. That's' why the perspectives of creation of Internet economy are largely predetermined by influence of social factors—readiness of economic agents to take additional risks for larger profit and tendencies in the sphere of the latest technologies.

Thus, a fundamental foundation and preconditions for further creation of Internet economy are created by the innovational economy which formed a society that accepts novelties, which include Internet entrepreneurship, and technologies necessary for solving the problems of Internet economy security.

Chapter 4
Modern Foundations of Internet Economy

4.1 Basic Principles of Modern Internet Economy

The sense of Internet economy is seen in its basic principles that determine its advantages and are the criteria of an economic systems' being assigned to Internet economy. One of such principles is freedom of entrepreneurial activities. It is a guarantee of growth of business activities that are a basis of Internet economy. This principle supposes bringing down the level of state interference with economic activities of economic agents to the minimum and provision of full action of the market mechanism.

Observation of this principle supposes creating and supporting favorable conditions for development of entrepreneurship—i.e., business climate, including competitive tax rates for business, accessibility of state services, competitive environment at sectorial markets (established by effective anti-monopoly policy), and lack of "institutional traps" that hinder the achievement of high effectiveness of business processes (e.g., bureaucracy and corruption).

Another principle is domination and active development of small entrepreneurship. Due to its specific features—high flexibility, adaptability, and inclination to implementing innovational products and business processes, small business, among all types of entrepreneurship, is a favorable environment for distribution of Internet technologies and implementing a new Internet format of doing business that supposes the maximum use of the Internet.

It should be noted that stimulation of development of small entrepreneurship should be done not by means of provision of certain preferences that inevitable leads to restraining other forms of entrepreneurship that perform their important roles in the process of functioning of Internet economy, but by means of supporting low barriers of entering in the sectorial markets. This will allow achieving high level of development of small entrepreneurship with formation of its other forms—i.e., preservation of the balance of business activity.

A.P. Sukhodolov et al., *Internet Economy vs Classic Economy: Struggle of Contradictions*, Studies in Computational Intelligence 714,
DOI 10.1007/978-3-319-60273-8_4

Another principle of modern Internet economy is domination of the service sphere (presumably trade) in the structure of GDP, i.e., development of economic system according to the post-industrial type. It is important, as maximal Internetization of business could be reached at the stage of "distribution" of the chain of value added. It should be noted that economic relations that are based on the Internet are applied to the sphere of B2C and B2B—that is, Internet business can cover not only sales of finished products to final consumers but supply of resources and final products to business partners (other companies).

However, preservation of sustainability of development of Internet economy requires strong foundation in the form of real sector of economy. That's why post-industrialization should not be viewed as a course of elimination of industry and agriculture but treated as a landmark for growth of business activity in the service sphere with preservation of usual volume of other spheres of economic activities.

The viewed principles include also domination of Internet deals and transformation of transaction relations in the favor of their Internetization. Internet deal is a deal between seller and buyer on the Internet—which performs the role of the market. That is, classic deals that suppose direct personal contact of seller and buyer and payment and transfer of a product in the moment of conclusion of deals with the help of cash within specific territorial space (office, store, etc.) should go to the background and be brought down to zero.

At that, Internet deals that suppose either full absence of buyer-seller contact or their interaction on the Internet (through social networks, specialized platforms, or seller's web-site), online payment, or delayed receipt of the product by the buyer (determined by the selected method and term of shipment) should be performed more often and become more popular in the sectorial markets.

These principles include also minimization of transaction costs. Due to the fact that most companies have their own web-site on the Internet, and most products and services are accessible for view and purchase on the Internet, a high level of consumer's knowledge on the market situation is achieved—which simplifies the process of decision making and, according to the classic economic theory, is the most important condition of their rationality.

As description of products and services is available on Internet web-sites, there's no necessity for long discussions of potential purchases and interaction between seller and buyer. Conclusion of Internet deals supposes reservation and/or payment for a product on the seller's Internet page and selection of the preferable means and term of shipment. All of this is done by the seller in a one-sided manner without direct participation of the seller who is just informed on the concluded deal as a result of these manipulations.

It is necessary to pay attention to such principle of modern Internet economy as achievement of the optimal ratio of demand and offer in sectorial markets. As in case of Internet deals the seller does not have a necessity to show specific products to buyer, for it suffices to post their images and detailed description on the Internet page, he does not have surplus of unsold (undemanded) products. Due to the

Internet format of doing business, a seller can purchase from supplier the exact volume of products that he has to sell to buyers.

In his turn, supplier can manufacture the required volume of products. A key moment here is buyer's readiness to wait until the selected (purchased) product will be manufactured and shipped. Therefore, the initial point of economic activities is demand, not offer. Internet business eliminates the necessity to rent or buy the trading premises with showrooms and samples of the offered products, hire and pay services of employees, etc. That is, Internet economy stimulates rationalization of economic activities and practical realization of the concept of economic effectiveness (establishment of market balance) of the classic economic theory.

Another principle of modern Internet economy is availability of the corresponding infrastructure that is based on the developed system of online payments and the system of Internet navigation. The system of online payments is necessary for online payment for purchases performed on the Internet. In includes various payment tools, the vivid examples of which are banking cards, WebMoney, LiqPay, Yandex Money, etc. These tools should be widely accessible and popular.

An important condition of successful work of Internet business is good navigation on the web-site of an Internet shop and in the sectorial market on the whole (through specialized platforms at which it is possible to compare prices and products and study existing offer in the market). Such navigation supposes quick search based on a lot of criteria set by the user. It could be in required not only by a buyer for selecting products but also by companies for determining their competitiveness in the market.

Let us mention another principle of modern Internet economy—active involvement of economic system into the process of international globalization and integration in the world economy. An important feature of Internet economy is its openness, which is necessary for it not to be closed and for it to develop. Participation in globalization and integration processes supposes cooperation with various international organizations, including joining the WTO and other regional integration unions.

This will allow eliminating national differences and unifying standards of doing business and stimulating global competition in national sectorial markets, which is important for provision of their development, and will create a chance for expansion of sales markets for domestic products abroad, which, in its turn, will accelerate the rate of economic growth and well-being of economic system and will raise living standards of the population.

And the last, but not least, is such principle of modern Internet economy as internationalization of economic activities and high share of international business activity in the structure of entrepreneurship. High share of foreign economic activities in economy is achieved by means of elimination of geographical boundaries of business with the help of its transfer into the Internet, where its products are accessible anytime and anyplace in the world.

Internet economy supposes absence of clear geographical orientation of business and its global coverage—i.e., active development of transnational entrepreneurship. At that, emphasis is made on development of international

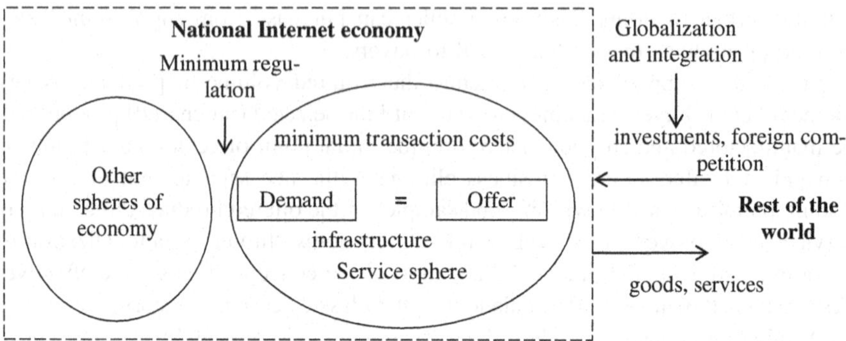

Fig. 4.1 Basic principles of modern Internet economy

transport and logistical infrastructure and the system of products shipment all over the world—which enable conclusion of international Internet deals.

Based on the above, the main principles of modern Internet economy could be presented in the following form (Fig. 4.1).

4.2 Internet Economy and Theory of Reproduction: A Newly Created Tandem

Let us try to study the system of public reproduction with participation of the corresponding information and technological sector with the help of a well-known methodology developed by K. Marx. A serious cause for consideration of this system is the idea that it is going to contain commodity and money exchange based on activity of a large volume of production sector within which limited resources are used; the issue of economic choice will stay urgent for a long time.

On the one hand, all public products and production, being in a situation of intense development of Internet economy, entered the stage of their division into two full parts:

- information sphere, where intellectual products are created that include Internet products. At that, Internet economy is an element of information sector;
- material sector, within which the issue of usual products and services takes place that are limited and figure as components constituting the foundation of commodity and money exchange.

It is clear that the natural and material form of total public product consists of multitude of information and traditional products. In a quantitative expression, it is represented by the sum of their costs.

On the other hand, the cost of the product received as a result of functioning of each of the above sub-levels is also divided into two components. At the first sub-level, their role belongs to the cost of information product necessary for receipt of the product in the information economic sector and the cost of information product and technologies applied in the processes of material production. The second sub-level includes the cost of material product necessary for realization of effective activities of the information sector and the cost of material product that poses special importance for normal existence of real sector of economy.

Each of the above elements cannot "work" without the viewed components in the conditions of evolution of the "new" economy. Intellectual labor needs real means of production (from computers to cameras), and its products need material carriers. In its turn, development of the sphere of material production won't take place without involvement of high technologies that are manifested in an information product.

The first branch consists of polythematic totality of the spheres involved in production and distribution of information products. The role of the Internet is very high here. Thus, more and more commodity items that are the results of intellectual labor become popular due to business actors' active use of the Internet—due to which such positions have qualities of Internet products.

The second branch contains the remaining spheres of national economy which, within the limits of creation of classic products and services, use the results of intellectual activities present in an information product.

Effective work of reproduction system in scale of the whole society requires a well-developed scheme of realization of commodities exchange between its two branches. Due to that, intellectual product, received in the first branch, acquires its material form that is important for future period of activity of this economic sector. Apart from that, it is necessary to note that such exchange allows the equivalent of the cost of second branch's product to be realized in specific information products that participate in production of usual goods and services.

For the purpose of the research, various proportional dependencies are not important, as they depend on the level of development of information sector of economy in a certain country. Let us assume that half of commodity cost of the first sub-level is paid beforehand by entrepreneurs and hired employees for purchase of products of the second sub-level, and another half is addressed for replenishment of own production means and consumption. As a result of this process, cost equivalent of one of the two parts of information product realizes its return to the first sub-level in the form of material production means and objects of consumption that appeared from the second sub-level where conditions are different. In exchange for a certain product of material production, owners and their hired help can receive an information product for application for consumer and production purposes.

Mutual profit of such exchange is ensured by means of money turnover, which allows achieving equal shares of the information product not limited by resource factors and limited product of material production. As replenishment of material production means and, therefore, satisfaction of material needs of workers of information and technology sphere of economy are necessary, creative product has to show itself in the money form and be exchanged for a product of material production. Such turnover could include a lot of financial and economic operations conducted by employees and their employers of the both sub-levels in the path leading to satisfaction of their production and personal needs. As we see, money in this case in equivalent due to which the market mechanism can support proportions between both economic sectors.

It is possible to conclude: money in any form of institutional agreement between individuals will function for a long time, as material and non-material goods and resources will exist for many a century.

Let us pay attention to the following situations: simple and expanded reproduction of total public product with participation of two sub-levels. Considering the procedure of simple reproduction leads to the following conclusions:

- new cost in the form of information product, received as a result of year-long labor of entrepreneurs and their employees at the first sub-level should be equaled to cost of a usual product, manufactured by those employed in the real production sphere;
- proportional dependence, within which each of sub-levels is divided into two parts, is related to the level of development of post-industrial sector of economy. In the countries with low level of development of this sector there is non-equivalent exchange between two sub-levels, as small volume of information product is comparable by cost to a large value of the number of products manufactured in the real sector of economy;
- substantiating the difference between the levels of development of two sub-levels of public reproduction could be performed on the basis of analysis of existing difference in capitalization of companies, norm of entrepreneurial profit, and payment for labor of these sub-levels' employees.

As could be guessed, the main goal of simple reproduction is consumption, but profit might play the role of stimulator of entrepreneurial activity at any moment. However, whatever its volume, this category will be only a basis for individual consumption of subjects of entrepreneurial activities. Simple reproduction is a large part of yearly reproduction in the expanded scale, and personal consumption preserves its value, staying with the motive of enrichment as one of the conditions of intense economic development.

With increase of the scale of information sector of economy in the conditions of stable existence of various factors, including the above, a large share of elements constituting total public product are compensated by limited material products from the second sub-level. Execution of such scenario with simple reproduction would lead either quick reduction of the level of prices for intellectual products or increase of prices for usual products, or to reduction of the share of material products useful for supporting the effective work of real production. Reduction of the level of prices for information products would lead to reduction of their issue. Increase of prices for usual products would lead to increase of their production. Reduction of the share of material products important for existence of the second sub-level would lead to reduction of the volume of applied main capital and reduction of the volumes of the issued products. Due to controversy, both are unacceptable for application within our model.

In order to decrease the influence of these contradictions, it is expedient to view the problem of functioning of expanded reproduction, the main components of effective realization of which should be excess of production of information product that is necessary for the second sub-level over the product of real sector of economy applied at the first sub-level.

Modern economy goes through quick development of the information sector which creates added value that exceeds total volume of the product that accounts for the beginning of the year.

Growth of cost of information product is easily distributed with a certain proportional dependence between its components: cost information product necessary for efficient work of information sector of economy and cost of the information product participating on the sphere of real production. Let us pay attention: at that, it is possible to see not only the monetary but also natural and material change of the specified parameters, which is explained by the need for growing information sector in expansion of its material component: production means and carriers of information.

It is clear that the products of the first branch which cost and number grew require the corresponding growth of the volume of product issued at the second sub-level of public. Thus, the first sub-level aims at taking the added product to the second sub-level. It its turn, the second sub-level strives for reacting by increase of production of own product that is important for realization of exchange procedures with the first sub-level. Based on the above, it is possible to formulate the economic law of exceeding development of information sector that is effective in the situation of transition to the post-industrial stage of development of economic system.

Such law of public reproduction is substantiated in the following:

- unlimited character of resources of information economy as a key stimulator of economic development in future;
- exceeding growth of labor efficiency in information sector of economy;
- stabilization of production volumes within material sector of economy.

4.3 Crisis in the System of Internet Economy: Reasons and Consequences

Generally speaking, the objects viewed above are not the only representatives of the products realized on the Internet. There are shares of Internet companies, artificially created online currencies, etc. Thus, if the category of Internet money is a usual thing with the progressive part of the world, the phenomenon of shares of online companies being traded in the stock market caused multiple arguments. Let us dwell in this issue in detail.

It is well-known that the main components of Internet economy—online business—is a young structure, due to which the methods of forecasting of its monetary fate are far from perfection due to practical lack of important experience. At that, Internet entrepreneurship is developing: more sustainable business models form and develop.

At present, most analysts think that formation of the cost of Internet assets differs from creation of such within traditional offline businesses. It is obvious that Internet projects can bring huge profit. But it is necessary to be careful even with such prospective project. So from which side should one approach the issue of evaluation of risks and perspectives?

Some authors think that within evaluation of Internet organizations it is important to take into account the indicators of their cost and specific indicators that are peculiar for a certain sphere of functioning of the viewed company. Three methods of evaluation are considered traditional: profitable, comparative, and estimate. The first one supposes assessment of the money flow of an online company in future and the following recalculation according to their cost at present. The second approach supposes calculation of company's perspectives on the basis of comparing separate characteristics of several similar companies. These are ratio of market cost to factual income of the company, ratio of company's cost to sales, etc. The supporters of the third approach prefer calculating the difference between real cost of all assets and real cost of debt liabilities of the company. A good addition to these calculations could be accounting of such indicators as the volume of the web-site's audience, accompanied by comparing the unique visitors and the total number of the web-site's users, as the online platform's loyal audience means existence of real possibilities for advertising (Sakoyan 2014).

Despite the above, evaluation of Internet projects is still built on the principles of intuition. Thus, more than half of assessment techniques require addition information for reasons of clarity. Intensively developing online companies face such problem, as the difference between their experience and profits that could be achieved in perspectives might be too big. Thus, there are no maximally convenient means of conduct of evaluation of Internet businesses as of now. It would be good to improve such state of affairs.

Cassidy made an effort to determine the most "expensive" Internet project of modern times, dwelling on the comparative approach. He was able to gather data on the largest online companies provided by Yahoo Finance and perform the comparison of a range of indicators:

- ratio of market price to factual income for the past 12 months;
- ratio of market price to inventory cost;
- ratio of the first indicator to supposed rates of growth of income divided by the ratio of market cost of the company to money flow.

The results of this comparison show that the least adequate market evaluation is peculiar for Twitter and Internet radio Pandora. Indeed, with low—even negative—

indicator of profitability—these project still work. The web-sites with positive market evaluation include Amazon and LinkedIn.

However, it is necessary to note that the conclusion on high evaluation of Twitter is based on the approach that is mostly used with traditional businesses. That's why this approach is less efficient as to the viewed projects.

Continuing the consideration of the problem, it is necessary to note: the authors interested in the issue of cost of Internet assets usually return to the so called bubble of dot coms which grew in late 1990s, reached its peak in 2000, and disappeared in 2001. That time was peculiar for establishment and quick development of commercial Internet projects which were expected to show large scale. Strangely enough, they became the issue of discussions on "new economy". Self-elimination of the bubble led to bankruptcy of thousands of Internet companies that became popular during the peak of its functioning. There were also legal claims to business actors accused of spending shareholders' money. However, such giants as Google, Amazon.com and eBay preserved their positions and grew.

The process of bubbles appearance starts with emergence of any perspective invention. For example, a browser with a simple name "Mosaic", which appeared in 1993, transformed the Internet into the structure represented by totality of interconnected pages. Creation of web-sites and their following use became simple and interesting processes. The lobbyists calling upon the society for deregulation of the market and implementation of a range of tax remissions became very active. At the end of 1995, the Internet allows the users to receive large profits from realization of various deals. Four years later, the conditions for development of commercial relations on the Internet became even better: the moratorium on sales tax was announced. As a matter of fact, legal changes do not provoke emergence of bubbles but, as practice shows, existence of the latter is impossible without the former.

The time came when independent people attached the Internet for the purpose of easy profit. Still, what is quick cannot always be safe. The logic and lessons of the past were forgotten.

Where there is deregulation, there's need for investments. The method of organization of financial provision for each bubble is unique, while the final goal is universal: the system should be able to accumulate unlimited volumes of money assets, creating expensive securities. At first, new online project created a field for venture investing. Later, a small number of startups brought their owners large income. It may look like a vicious circle: the profits received as a result of investing are transferred into new funds of venture financing and then to other projects interesting for new investors who invest large sums into them. Thus, the circle is finished with venture funds again.

At one time, mass media has a lot of information on young people who made their first millions on the Internet. Real gaps of a new system of online business were not interesting for a large audience. Congressmen, who were able to influence the regulating agencies, were never against the tide of "tax" money.

The main care of online projects' top managers was maximization of profits and realization of efforts for preserving the companies against the unexpected losses in the moment of the bubble's destruction. In March 2000, the first sign of possible threats appeared. Representatives of the largest investment establishments advised that the online projects do not enter the stock market in April. These recommendations meant that the market had reached the stage of saturation. In the several following months thousands of investing subjects that had become the owners of unrealized income began selling their shares for the purpose of future payment of tax liabilities. The turn for panic came. The bubble exploded.

Fictitious contents of the bubble disappears when the quick formation of false hopes of vendors is followed by their quick downfall. Despite the fact that the direct participants of these events constituted a small share of companies, the recession of the market influenced the economy on the whole, provoking the recession in early 2001. Society faced the following question: how is it possible to compensate $8 trillion of fictitious cost created by the bubble?

Thus, only the flow of capitals is required for creation of a bubble. It starts cracking due to existence of securitized debts.

In our understanding, securitization means the issue of securities based on several existing ones for the purpose of more predictable and less risky object of investing than the sum of elements that forms it. Large volumes of such securities, supported by credits, allows banking structures to be intermediaries, for even in case of bankruptcy of a certain debtor payments from other members of the relations do not stop, which means that the bank that performs realization of securitizes debts does not suffer or experiences less inconveniences than it would in the situation when it would issue the credits directly. From theoretical point of view, the risks of the bank could be distributed for financial markets among experienced institutional and investment actors.

Let us conclude: there are not threats to market cost of Internet companies. Taking into account that Internet projects are projects of the future, it is necessary to rely on the future. No one denies emergence of a new bubble. In view of the current state of our economy, the worst that could happen is its absence (Janssen 2008).

Chapter 5
Methodological Aspects of Study of Internet Economy

5.1 Algorithm of Creation of Internet Economy

Despite the key role of market self-management in existing Internet economy, the main load is set on the state. While in certain countries, which economic systems' necessary preconditions formed by themselves, Internet economy formed spontaneously, the process of its development in most countries is a manageable process and requires active interference of the state. The algorithm of creation of Internet economy is presented in Fig. 5.1.

As is seen from Fig. 5.1, the process of Internet economy creation has four successive stages. The first stage supposes creation of institutional platform for Internet economy. This platform includes the following important structural elements. Firstly, the institute of normative and legal provision of creation and conduct of Internet business. It has to clearly specify the notion of Internet entrepreneurship, rights and liabilities of Internet entrepreneur, and the main aspects of his activities, including the procedure of registration and closure of Internet company.

Secondly, the institute of regulation of transaction relations of economic агентов on the Internet. Here the main attention is paid to rights and liabilities of a buyer during conclusion of Internet deals, peculiarities of their registration, regulation of the main types of such deals, and the order of accommodation of disputes during their conclusion, execution, and termination. In particular, it is necessary to pay attention and set in the law the terms for return of products purchased on the Internet, and seller's responsibility for quality and terms of shipment of such products, procedure of return of paid money so it is convenient for buyer—otherwise it would reduce attractiveness of Internet deals as compared to classic deals.

Thirdly, the institute of regulation of international Internet deals. As Internet economy is largely oriented at internationalization of entrepreneurial activities, its foundations should be determined, and interests of domestic Internet entrepreneurs should be protected by law. A transnational character of these deals predetermines

A.P. Sukhodolov et al., *Internet Economy vs Classic Economy: Struggle of Contradictions*, Studies in Computational Intelligence 714,
DOI 10.1007/978-3-319-60273-8_5

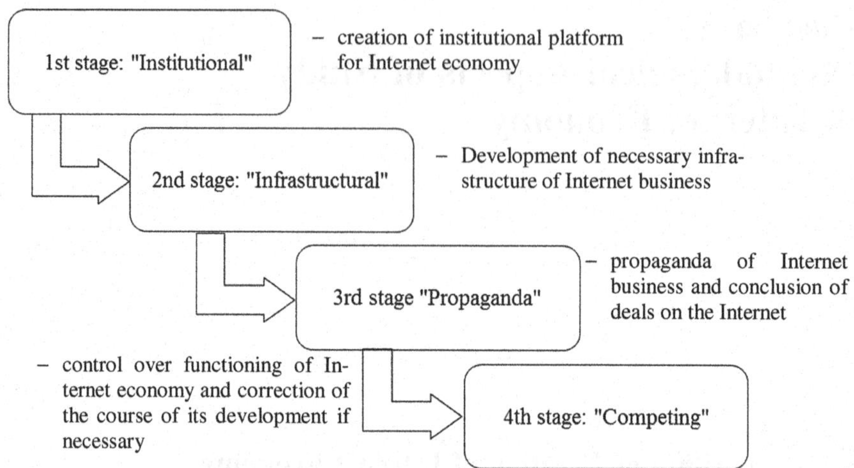

Fig. 5.1 Algorithm of Internet economy creation

the necessity for involving authoritative and acknowledged international organizations, conclusion of deals and management of relations with which is the most important task of the state at this stage.

Creation of the institutional platform has to reduce the level of uncertainty and risk of doing Internet business and conclusion of deals on the Internet, this increasing their attractiveness for economic subjects. This platform determines "rules of the game" for participants of economic relations on the Internet. It should be transparent and clear for all members of such relations, as well as stable—to keep their trust in Internet deals.

The second stage supposes creation and development of necessary infrastructure (apart from the institutional infrastructure that was created at a previous stage): technological, financial, HR, and transport and logistical. The technological infrastructure of Internet entrepreneurship includes necessary means, equipment, and technologies. In particular, it is access to the Internet, software, etc.

Financial infrastructure of Internet entrepreneurship is development of the system of online payments, system of insurance of Internet deals' risks, and system of crediting (including lease) and investing of Internet entrepreneurship. These financial systems are to provide Internet entrepreneurship with necessary financial resources for successful functioning and development, raise its sustainability and provide it with reliable, safe, and modern payment tools necessary for Internet deals.

HR infrastructure of Internet entrepreneurship is preparation of highly-qualified specialists for working with Internet entrepreneurship, capable of development and implementation of innovations in the sphere of information and communication technologies for supporting its competitiveness. This supposes modernization of the sphere of science and education in view of actual needs of Internet entrepreneurship.

Transport and logistical infrastructure of Internet entrepreneurship should allow for quick and reliable supply of products purchased via the Internet. This infrastructure should unite geographically distant territories, differentiate transport ways and means available for them, and provide high speed of movement on the territory of the country. It should correspond to the newest international standards.

The third stage is related to propaganda of Internet entrepreneurship and conclusion of deals on the Internet. When all necessary conditions for formation of Internet economy are created, it is necessary to start (activate) the process of its creation. In some cases, this may not be necessary—if the society is ready for conclusion of Internet deals—however, as a rule, this stage is necessary. The propaganda is conducted with the help of social advertising and teaching courses and includes the following directions.

First direction: preparation of economic agents to conduct of Internet entrepreneurship and conclusion of Internet deals. It includes teaching the population the basic work on the Internet and use of computer. Such training might be necessary for some people (e.g., the retired) or for all groups of population—depending on the level of society's informatization and level of its development.

Second direction: popularization of Internet entrepreneurship. It is held in the form of consultations of experts and representatives of various bodies of state power with active entrepreneurs and is aimed at establishing the dialog between them and attracting interest to Internet entrepreneurship. It could be stimulated for the purpose of provision of tax or credit preferences, which have to be cancelled later—when the interest in Internet entrepreneurship will become natural.

The third direction: popularization of Internet deals. It is aimed at buyers and should answer their questions regarding reliability of the deals and their attractiveness—and maybe inform of the possibility of conclusion of the deals. There are no limitations as to characteristics of such buyers: the basics of conclusion of Internet deals should be taught to pupils and the retired, urban and rural residents alike, etc.

The fourth stage includes control over functioning of Internet economy and correction of the course of its development if necessary. Initially, when the interest to Internet deals just appeared and Internet entrepreneurship was treated as something new, with mistrust and care, it was important to support this interest and support entrepreneurial initiatives. A large share of state interference with the process of formation of Internet economy is still preserved.

The role of state consists in provision of correct work of the announced basic principles of Internet economy and full realization of current potential of economic system in creation of Internet economy. It should observe the actions of all market agents, coordinate them, and direct them if necessary.

When Internet entrepreneurship became an everyday aspect of economic activities of the economic system and Internet deals became a usual practice, the need for state regulation reduces and the level of its involvement into functioning of the Internet were minimized. The main attention should be paid to crisis management of Internet economy, related to increase of its sustainability. The state set the general tone and determined national political course for development of Internet economy.

The process of development of Internet economy may take a long time, and its success depends on the presence of corresponding preconditions—as, despite the manageability of this process, forced formation of Internet economy cannot ensure effective work of its participants and achievement of expected advantages. That's why, initially, (before realization of the first stage) it is necessary to evaluate the possibility of creation of Internet economy and the economic system's readiness for that.

5.2 Methodology of Managing Internet Economy

As was said above, Internet economy supposes self-organization and market self-management with minimum share of state participation. Still, the necessity for its regulation appears due to two interconnected reasons. The first one is growth of cybercrimes that pose serious threat for Internet economy. As most deals of economic subjects in Internet economy are performed on the Internet, they are vulnerable to hackers, i.e., cyber criminals, who can hack sellers and buyers' accounts, change the results of market rankings, etc.

Security of Internet entrepreneurship and Internet deals is a basis of stability of Internet economy. That's why fighting cybercrimes is an important task set on the state due to its being among the number of public (non-profit) goods. Without the corresponding measures, Internet economy can find itself in the hands of cyber criminals and become unmanageable for the state. In this case, it would be difficult or even impossible to preserve its high effectiveness and restore its market balance and state control over the situation.

The second reason is unstable development of Internet economy and high level of its aptitude to economic crises. Even without purposed negative influence of internal and external forces, Internet economy is subject to crises. Firstly, high level of freedom of financial markets and actions of financial speculators (investors in these markets) predetermines the lack of their clear connection to the situation in the real sector of economy, high volatility, and inclination to emergence of "financial bubbles".

Secondly, domination of small entrepreneurship in Internet economy determines total low corporate responsibility of business and instability of its development, related to high probability and frequency of market re-orientation and bankruptcy of small companies. Thirdly, orientation of Internet economy at constant innovational development is a reason for high risk components related to unpredictability of consequences of development and implementation of innovations. Therefore, the state's efforts should be aimed for constant crisis management of Internet economy.

Let us view the most perspective methods of Internet economy management. One of them is tax regulation, conducted within state fiscal policy. Depending on current needs of market and society, tax could perform stimulating or restraining role. Thus, provision of tax preferences to Internet companies that correspond to certain criteria, established according to the selected course of development of Internet economy—say, high innovational activity, active support for increase of the level of population employment, corporate responsibility, etc.—can direct the behavior of market players in the corresponding course.

Another method is regulation of inflation, performed within state monetary policy. As Internet companies are not connected to national companies and want to enter the global arena, currency rate influences the level of their competitiveness in the global markets, providing a pricing advantage as compared to foreign rivals or even determining higher prices for their products.

That's why in the process of regulation of the course of national currency of Internet economy, the state should take into account interests of consumers interested in reduction of inflation for increase of living standards of the population and of national Internet entrepreneurship, interested in growth of inflation for increase of their global competitiveness, and should find and support balance of these interests.

Another method of management of Internet economy is formation and development of necessary institutes within the state institutional policy. As a strong institutional basis lies in the foundation and development of Internet economy, the state should support it with all possible means. With appearance of new technologies it is necessary to institutionalize them—i.e., legally establish the notion and possibilities (and limitations) of their application.

With Internet economy, it is inadmissible to allow for serious delay of normative and legal basis as to economic reality, as it will be a restraining factor on the path of creation and implementation of innovations, while quick technological progress is one of the most important conditions of establishment of Internet economy. That's why the state should observe the changes in the market and increase the institutional platform.

The methods of Internet economy management include creation and development of the corresponding infrastructure, realized through the state infrastructural policy. Modern infrastructural provision is necessary for successful functioning of any business, including for Internet entrepreneurship, which is a core of Internet economy.

At present, the leading and still developing tool of development and effective management of infrastructure entrepreneurship is the system of public-private partnership. It allows reducing financial load on state budget and attractive private investments into realization of infrastructural projects. Obviously, the efforts of state in Internet economy should be aimed at development of this system within its infrastructural policy.

The methods of management of Internet economy include also stimulation of innovative activity, performed with the help of state innovational policy. The tools of such stimulation could be co-financing or provision of grants for scientific research, support for development of international educational and scientific communications and cooperation, and determination of national priorities of development of science and education in view of actual needs of society and economy.

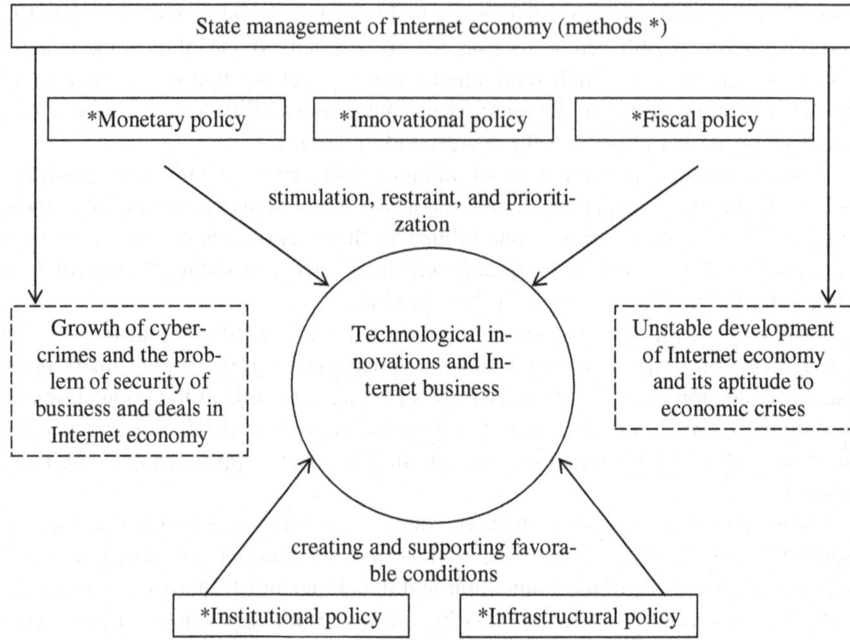

Fig. 5.2 Methodology of management of Internet economy

The performed analysis allows providing the methodology of management of Internet economy in the following way (Fig. 5.2).

5.3 Approaches to Evaluation of Internet Economy Effectiveness

Evaluation of effectiveness supposes comparison of profits and expenses for their achievement. As to Internet economy, it is possible to distinguish its four main advantages as compared to the classic economy: higher rationality, smaller transaction costs, larger innovative activity, and bringing state management down to minimum—which determine corresponding approaches to evaluation of its effectiveness.

The approach based on analysis of the rationality level supposes the use of the following formula for evaluating effectiveness of Internet economy:

$$E_{ie} = |D_{ce} - S| - |D_{ie} - S| / |E_{ic} - E_{cc}| \tag{5.1}$$

where

E_{ie} effectiveness of Internet economy;
D_{ie} total demand for all goods and services in economy;
S total offer of products and services in Internet economy;
S total offer of products and services in classic economy;
E_{ic} total expenses of Internet companies required for satisfaction of current demand;
E_{cc} total expenses of classic companies required for satisfaction of current demand.

In formula (5.1) the level of rationality of economy (numerator) is assessed through comparison of demand and offer—the difference between then in classic and Internet economy shows advantages of the latter. In this case, it is not important what is larger—demand or offer—so both are taken by the module.

Then, their ratio to expenses necessary for achievement of the determined additional rationality is calculated, it is determined through finding the difference between total expenses of Internet companies required for satisfaction of current demand and total expenses of classic companies that would have been required for its satisfaction in classic economy.

Within the approach that emphasizes on transaction costs, the formula for assessment of effectiveness of Internet economy could be presented in the following way:

$$E_{ie} = S_{ie}/|TE_{ic} - TE_{cc}| \tag{5.2}$$

where

S_{ie} total offer of products and services in Internet economy;
TE_{ic} total transaction expenses of Internet companies and consumers;
TE_{cc} total transaction expenses of classic companies and consumers.

While within the first approach the main attention is paid to the numerator (profit) that is supposed to reach its maximum within Internet economy, within this approach the denominator comes into foreground (expenses for its receipt), which is brought down to minimum for growth effectiveness. At that, efficiency of Internet companies (level of satisfaction of total public demand) has a secondary role and is not taken into account.

In formula (5.2), a key role in provision of effectiveness of Internet economy belongs to achievement of lower total transaction expenses of Internet companies and consumers as compared to total transaction expenses of classic companies and consumers in classic economy. During calculation of effectiveness, the ratio of total offer of products and services in Internet economy to the found difference in transaction expenses is calculated. The numerator is presented by volume of GDP, volume of deals, or total added value.

In the context of the approach that concentrates on innovational activity, assessment of Internet economy effectiveness is performed by the following formula:

$$E_{ie} = |IA_{ic} - IA_{cc}|/ET_{ia} \quad (5.3)$$

where

IA_{ic} innovational activity of Internet companies;
IA_{cc} innovational activity of classic companies;
ET_{ia} total expenses of society (state) for stimulation of innovational activity in economy.

In formula (5.3), attention is paid to numerator which has to strive to the maximum for achievement of high effectiveness of Internet economy. As in Internet economy high innovational activity of its main economic subjects—Internet companies—is achieved with market mechanism, expenses of the public (state) for its stimulation should be minimal.

That's why during evaluation of effectiveness of Internet economy innovational activity of Internet companies in Internet economy is compared to innovational activity of classic companies in classic economy. The received difference is compared to total expenses of the public (state) for stimulation of innovational activity in economy.

Innovational activity could be evaluated as the volume of developed leading production technologies for a certain period, volume of frequency of implementation of such technologies for modernization of business processes or volume of manufactured innovational products and services. Selection of a specific indicator depends on accessibility of statistical information. At that, it is important to state the selected level of innovativeness, which could be absolute (new for the world economy on the whole, completely new) or relative (new for this company, sphere, or national economy).

The approach that emphasizes the level of state interference with economy, the formula for evaluating the effectiveness of Internet economy will have the following form:

$$E_{ie} = EE/|ESR_{ie} - ESR_{cc}| \quad (5.4)$$

where

EE economy's efficiency;
ESR_{ie} expenses for state regulation of Internet economy;
ESR_{cc} expenses for state regulation of classic economy.

In formula (5.4), efficiency of economy could include one, several, or all such indicators as offer of products and services, innovational activity of enterprises, volume of GDP, volume of deals, or total added value. Main attention is paid on numerator, where the difference between expenses for state regulation of Internet economy and such expense for regulation of classic economy is calculated.

It should be noted that formulas (5.1)–(5.4) are given in a generalized form and could be detailed depending on possibilities and goals of calculations. In our case, these formulas have to reflect the general sense of each of distinguished approaches so they could be compared. However, even during consideration of the given formulas in the generalized form it is possible to see the following.

Firstly, a lot of indicators that are used in them are difficult to calculate, and they cannot be taken in a final form from existing statistical reports. Moreover, bringing the numerator and denominator to common measurement units is a difficult task. Secondly, each of the given formulae reflects only certain aspects of Internet economy effectiveness and does not allow building a comprehensive picture of it. That's why search for the means of combining existing formulae and creation of a complex approach to evaluation of Internet economy effectiveness is a perspective direction of further research.

Chapter 6
Peculiarities of Formation and Development of Internet Economy in Russia

6.1 Preconditions of Formation of Internet Economy in Russia

On the whole, general climate for formation of Internet economy in modern Russia could be characterized as rather favorable. This is determined by presence of the corresponding preconditions. Firstly, there is actively developing infrastructure for creation and conduct of Internet entrepreneurship. Thus, technologies of Internet business have appeared and actively develop—i.e., the developed technological infrastructure of Internet entrepreneurship is available.

Financial infrastructure of Internet entrepreneurship is rather developed in Russia. The issue of banking cards is an inseparable component of Russian banks' activity. The issue of special cards for payments on the Internet becomes more popular—such cards are copies of the main banking cards of a customer. This makes Internet deals more secure, as a customer does not risk his main card and gives away the information on the substitute card.

The number and popularity of electronic money systems grows; they perform the role of payment tools during execution of Internet deals. The most popular Russian systems are RBK Money (RUpay), Rapida (Rapida Online), and Ediny Koshelek (Single Wallet). HR infrastructure of Internet-entrepreneurship is also formed and presented by a large number of specialists in the sphere of information and communication technologies (primarily, programmers), economists, and managers who can manage an Internet company, bankers who can support the system of financial accounting and payments for Internet companies, etc.

Secondly, the institutional environment of Internet entrepreneurship and conclusion of Internet deals is already formed. According to the data as of January 1, 2017, there is a range of laws in Russia that regulate them. Federal law of the Russian Federation dated April 5, 2013 No. 44 "On the contract system in the sphere of purchases of products, works, and services for provision of state and municipal needs" (The State Duma of the RF 2013) and Federal law No. 94 dated

A.P. Sukhodolov et al., *Internet Economy vs Classic Economy: Struggle of Contradictions*, Studies in Computational Intelligence 714,
DOI 10.1007/978-3-319-60273-8_6

July 21, 2005 “On placement of orders for supply of products, execution of works, and provision of services for state and municipal needs” (The State Duma of the RF 2005) determine conditions and the order of placing state orders on the Internet.

The Decree of the Ministry of Economic Development of Russia No. 54 dated February 15, 2010 (Ministry of Economic Development of the RF 2010) determines the order of public auction in electronic form during sales of property (company) of debtors in the course of procedures applied in the case on bankruptcy, sets requirements to electronic platforms and operators of electronic platforms during public action in electronic form during selling the property (company) of debtors in the course of the procedures applied in bankruptcy cases, and sets the order of confirmation of correspondence of electronic platforms and operators of electronic platforms.

There is also a project of the Federal Law “On e-commerce” No. 11081-3 dated October 3, 2000 (The State Duma of the RF 2000) which reflects the main notions and settings of Internet entrepreneurship. Except for the above normative and legal documents devoted to creation of the institutional environment for development of Internet economy, there are a lot of similar documents indirectly related to certain aspects of Internet companies’ activities and conclusion of Internet deals. This allows stating the acceptable level of uncertainty and risk related to their conduct.

Thirdly, there are positive shifts in development of market situation of Internet trading. Thus, on the one hand, there is growth of demand for Internet products. The results of survey of Russian Internet stores’ customers, performed in 2015 (PricewaterhouseCoopers 2015), showed that customers’ trust to Internet deals is rather high and grows as of now—and they are ready to purchase goods and services on the Internet.

The number and cost of purchases on the Internet grow—which is confirmed by positive dynamics of total cost of Internet deals in Russia (Fig. 6.1).

As is seen from Fig. 6.1, while in 2000 the total cost of Internet deals in Russia constituted only $2 billion, it grew by $5 billion in 2005, and by $11 billion in 2010. In 2016, it constituted $19 billion, and the 2017 is expected to bring the level

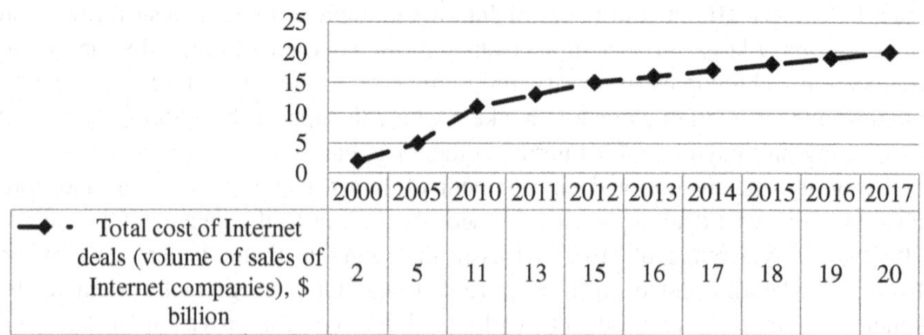

	2000	2005	2010	2011	2012	2013	2014	2015	2016	2017
Total cost of Internet deals (volume of sales of Internet companies), $ billion	2	5	11	13	15	16	17	18	19	20

Fig. 6.1 Dynamics of total cost of Internet deals in Russia in 2000–2016 and the forecast for 2017. Reproduced from Zubraeva (2015)

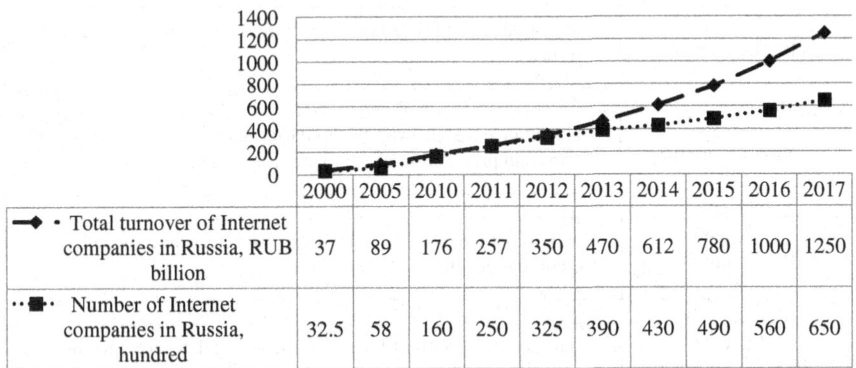

	2000	2005	2010	2011	2012	2013	2014	2015	2016	2017
Total turnover of Internet companies in Russia, RUB billion	37	89	176	257	350	470	612	780	1000	1250
Number of Internet companies in Russia, hundred	32.5	58	160	250	325	390	430	490	560	650

Fig. 6.2 Dynamics of indicators of development of Internet entrepreneurship in Russia in 2000–2016 and the forecast of total cost of Internet deals for 2017. Reproduced from InSales (2015)

of $20 billion. Its average annual growth has been exceeding 5% over the several recent years.

On the other hand, the offer grows—that is, the number of Internet companies becomes larger. This is confirmed by the annual increase of total turnover of Internet companies and their number in Russia. Dynamics of these indicators for the recent years is given in Fig. 6.2.

As is seen from Fig. 6.2, while in 2000 the total turnover of Russian Internet companies constituted RUB 37 billion, it grew to RUB 89 billion in 2005, to RUB 176 billion in 2010, and exceeded RUB 1000 billion in 2016. Average annual growth of this indicator exceeds 25%. In its turn, the number of Internet companies in Russia grows annually by 15%. Thus, there were only 32,500 of them in 2005, 56,000—in 2016, and the forecast for 2017 is 65,000.

The results of the performed analysis of preconditions of Internet economy formation in modern Russia show the presence of necessary conditions and economic system's readiness to its creation. However, successful conclusion of this process is largely predetermined by existence of objective necessity, demand for Internet economy by Russian market agents, and correspondence of the course of its establishment to national priorities and plans of strategic development and support for the global competitiveness of the country in the long-term.

6.2 Main Stages of Formation and Development of Internet Economy in Russia

In order to determine logic and perspectives of further development of the process of formation and development of Internet economy in modern Russia, let us view its main stages (Fig. 6.3).

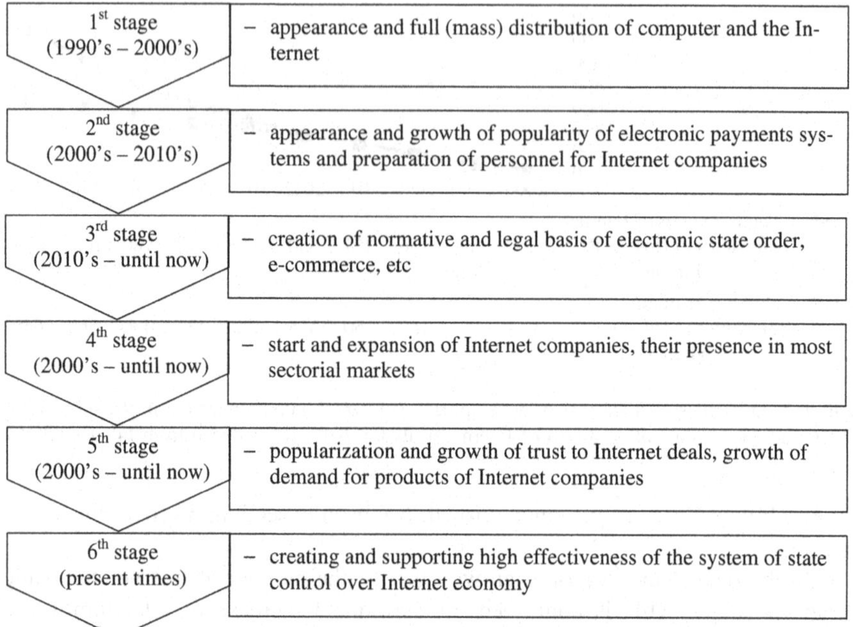

Fig. 6.3 The process of formation and development of Internet economy in Russia

As is seen from Fig. 6.3, these stages are conventional and not successive, but rather parallel. Despite general similarity to the above algorithm of creation of Internet economy, this process had its specifics in Russia. In particular, the infrastructural stage preceded the institutional one—that is, the institutional base was delayed as to development of the corresponding infrastructure, so incompleteness or lack of normative and legal basis was and is a restraining factor of formation and development of Internet economy in Russia.

The first stage was appearance and full (mass) distribution of computers and the Internet. The results of a sociological survey, performed in 73 rural and urban territories of Russia with 22,500 people, showed that as of 2016, 68% of able-bodied Russians were covered by the Internet. 57% of Russians, which is 66 million people, use the Internet on a daily basis (FOMnibus 2016).

In 2014, a course at development of the information society was established at the state level in Russia, and a corresponding state program was issued (Government of the RF 2014), which proclaimed mass computerization of society and provision of general access to the Internet among national priorities. At present, it is possible to state sufficient technical and technological equipment of Russia for development of Internet economy.

The second stage was peculiar for emergence and growth of popularity of electronic payments system and training of personnel for Internet companies. As of now, there are a lot of leading global systems of electronic payment in Russia, as well as many domestic systems. At the same time, the process of formation and development of financial infrastructure continues due to growth of society and business's needs, as well as quick development of technologies and the necessity for constant modernization of the systems of electronic payments in view of recent requirements of modern times.

Growth of popularity of Internet entrepreneurship and Internet deals increases the need for development of the corresponding HR infrastructure. As of 2016, there are 400,000 specialists in the sphere of information and communication technologies (highly-qualified programmers in Russia). Each year, 25,000 specialists graduate from educational establishments of the system of higher education. However, as practice shows, modern Russia requires twice as many such specialists in order to satisfy the growing need for them (Sokolov 2016).

The third stage of formation and development of Internet economy in Russia was related to creation of normative and legal basis of electronic state order, e-commerce, etc. Despite the fact that it is impossible to state full correspondence of such base to actual needs of economic subjects and its insufficiency for completion of the process of formation of Internet economy in Russia, there are clear efforts of the state to provide this correspondence and striving for modernization and active development of institutional provision of Internet deals and Internet entrepreneurship.

It is necessary to note presence and larger aggravation of the problem related to lack and low quality of transport and logistic infrastructure of Internet-entrepreneurship in Russia. Without effective system of supply, it won't be able to enter the global level or even cover domestic regional markets. Such infrastructure is present only in large regional centers, while most agricultural territories are hardly accessible for shipment of Internet products.

The fourth stage was peculiar for start and expansion of Internet companies and their presence in most sectorial markets. The most popular products over the recent years include such Internet products as electronics and household appliances (RUB 147 billion), clothes and footwear (RUB 93 billion), computer equipment (RUB 68 billion), car spare parts (RUB 38 billion), and mobile phones (RUB 32 billion).

There is close interconnection between the presence of the corresponding transport and logistical infrastructure and the level of development of other types of infrastructure and the volume of Internet deals. In 2016, they still dominate in large regional centers and lack in rural territories (InSales 2015).

The fifth stage included popularization and growth of trust to Internet deals and growth of demand for products of Internet companies. With strengthening of security measures of Internet deals and protection of information, the risk level and mistrust of Russian consumers to them reduce. At the same time, propaganda of these deals and Internet entrepreneurship is lacking. Due to that, despite the moral readiness to creation of Internet economy from entrepreneurs and consumers, lack of technical skills and necessary knowledge is a barrier in that.

The sixth stage supposes creating and supporting high effectiveness of the system of state control over Internet economy. Despite the inclination of Russian government to active regulation of most spheres of economic activities, Internet entrepreneurship and Internet deals are characterized by deregulation and almost full absence of necessary control, which makes them vulnerable to risks (cyber-crimes and economic crises).

Strengthening of social tension and attempts of Internet entrepreneurs to build dialog with public authorities bodies lead to attraction of its attention to the problem of monitoring and control over Internet deals, which is a positive tendency. However, the rate of positive shifts on this issue is still very low, so the problem is not yet solved. It is possible to suppose that after the end of this stage the process of Internet economy formation in Russia will approach its logical finish in view of parallel end of other stages mentioned above.

6.3 Scenarios of Internet Economy Development in Russia

Perspectives of formation and development of Internet economy in modern Russia are determined by socio-political factors, based on state and society's interest in its development, and by creation of necessary resources, which belongs to the sphere of economy. With the help of the theory of games and based on opinions of leading experts, who include (Dudina et al. 2017; Vendil Pallin 2017; Dubrovskaya and Kozhemyakin 2017; Gnezdova et al. 2016) and the authors' statements, we distinguished three possible scenarios of development of Internet economy in Russia, given in Table 6.1.

Table 6.1 Scenarios of development of Internet economy in Russia

Characteristics of scenarios	Scenarios		
	Optimistic	Realistic	Pessimistic
Actions of state and business	Timely strengthening of institutional environment and development of necessary infrastructure	Delay of institutional environment and development of necessary infrastructure as to needs of market	Large underrun of institutional environment and development of necessary infrastructure
Society's readiness	Quick growth of consumers to Internet deals	Slow growth of consumers' trust to Internet deals	Mistrust of consumers to Internet deals
Expenses for creation of Internet economy	$350 billion	$175 billion	$85 billion
Result	Quick formation of Internet economy, advantage of $1200 billion	Long process of formation of Internet economy, advantage of $500 billion	Incompletion of the process of formation of Internet economy, advantage of $350 billion
Probability	0.25	0.45	0.30

Based on the data from Table 6.1, let us determined the most profitable scenario for Russia. For that, let us perform corresponding calculations according to the following formula:

$$Prof_{scenario} = (Res_{scenario} - Exp_{scenario}) * p_{scenario} \quad (6.1)$$

where

$Prof_{scenario}$ profit from realization of a certain scenario;
$Res_{scenario}$ result of realization of a certain scenario;
$Exp_{scenario}$ expenses for realization of a certain scenario;
$p_{scenario}$ probability of realization of a certain scenario.

We have the following:

- $Prof_{optimist} = (1200 - 350) * 0.25 = \$212.5$ billion;
- $Prof_{realist} = (500 - 175) * 0.45 = \$146.2$ billion;
- $Prof_{pessimist} = (350 - 85) * 0.30 = \$79.5$ billion.

Obviously, the best ($212.5 billion) scenario is the optimistic one—despite its least probability. The most probable realistic scenario is rated second as to the level of profitability ($146.2 billion). The least profitable ($79.5 billion) is pessimistic scenario, characterized by average probability. Let us view them in detail.

The optimistic scenario supposes timely strengthening of the institutional environment and development of necessary infrastructure (for which the announced expenses are aimed) and quick growth of consumers' trust to Internet deals. It will ensure quick formation of Internet economy (over 10 years) and the advantage of $1200 billion. This advantage consists of increase of the volume of GDP (by means of increase of internal and international Internet deals), growth of competitiveness of national economic system and general growth of effectiveness of its functioning.

This scenario is oriented at high level of state and society's interest in creation of Internet economy, which supposes its active propaganda and the corresponding correction of national political course. It is necessary to note the necessity for favorable influence of external factors—stabilization of geopolitical situation necessary for providing Russia with a possibility to concentrate efforts on solving internal tasks of economic growth and development, as well as creating Internet economy in the countries that are close economic partners of Russia, which is necessary for provision of expediency of its formation in the national economic system.

According to the realistic scenario, we have delay of the institutional environment and development of necessary infrastructure as to market needs and slow growth of consumers' trust to Internet deals. It supposes long process of Internet economy formation (15–25 years) and advantage of $500 billion. This scenario is treated as the most probable, as it supposes preservation of the current status quo—i.e., minimal change of state policy and public opinion.

This scenario supposes that the obvious readiness of modern Russian society finds itself in multiple "institutional traps"—bureaucracy and corruption—and faces lack and low level of development of necessary infrastructure, primarily = transport and logistical. It is necessary to pay attention to the fact that these peculiarities of modern Russian economy are barriers not only for creation of Internet economy but for less radical modernization of economic system, as well as realization of any innovational initiatives.

The pessimistic scenario supposes strong underrun of the institutional environment and development of the necessary infrastructure, as well as consumers' mistrust to Internet deals. It will result in incompletion of the process of formation of Internet economy for 25 years and a small advantage of $350 billion. This is the least favorable scenario, as it supposes preservation of the current trajectory of economic growth of the Russian economic system, while most of developed countries (OECD) passed to a new quality of such growth and a new course of development of their economic systems.

A long delay in post-industrial level of economic development might lead to Russia's accepting a status of underdeveloped country in the aspect of socio-economic development—which contradicts her international interests and will not allows for realization of the designated policy in the global arena. Under the conditions of globalization, non-correspondence to actual global economic tendencies might lead to economic isolation and loss of support from other members of international economic relations. High probability of realization of this scenario is a serious reason for worries.

It should be emphasized that the performed scenario analysis and forecasting are based on a hypothesis of stability of internal and external economic situation in Russia and small influence of external factors on the process of formation and development of national Internet economy. The probability of the distinguished scenarios is determined by the method of expert evaluations on the basis of studying analytical information, which predetermines its low precision and relative character of these scenarios.

However, in view of dynamics of change of global markets' situation and instability of geo-political situation in the world economy, it is possible to state the possibility of emergence of additional scenarios of development in the situations that create new possibilities either for formation of Internet economy in Russia or for obstacles to this process. The detailed analysis of potential and real influence of these factors and compiling more precise and expanded forecasts of development of Russian Internet economy is a perspective direction for further scientific studies.

Conclusions

Thus, it is possible to conclude that despite preconditions for formation of Internet economy in modern Russia, this process faces serious institutional, infrastructural, and social barriers. In order to overcome them, we offer the following recommendations. Firstly, it is necessary to strengthen institutional environment of Internet entrepreneurship—in particular, to pass Federal law "On e-commerce", the project of which (No. 11081-3) has been under consideration since October 3, 2000.

We also deem it necessary to supplement the Article 3 "Definitions" of this law. In the current edition it includes only two participants of e-commerce: a person conducting e-commerce and a customer. However, two more subjects participate in Internet deals:

1. Financial intermediary—organization that provides services on conduct of payment for services of Internet company by its customers;
2. Supply intermediary—organization that provides services on shipment of products from Internet company to its customers.

This specification will allow eliminating the gap in terminological apparatus of Internet entrepreneurship and avoiding mistakes in treatment of actions and responsibilities of these subjects.

Secondly, it is necessary to ensure development of the necessary infrastructure. Emphasis should be made on transport and logistical infrastructure. In view of insufficiency of own financial resources with the state for creation and provision of correspondence of objects of transport and logistical infrastructure to actual needs of Internet entrepreneurship, it is expedient to attract private investors to this process on the basis of public-private partnership.

This form of cooperation allows preserving state control over development of transport infrastructure, expanding sources of its financing, and increasing effectiveness of management of projects on its development. The following recommendations are offered for increase of attractiveness of transport and logistical projects for private investors:

A.P. Sukhodolov et al., *Internet Economy vs Classic Economy: Struggle of Contradictions*, Studies in Computational Intelligence 714,
DOI 10.1007/978-3-319-60273-8

- provision of state guarantees of investments return;
- providing the investors with corporate tax exemptions of 3% during the whole period of project's payback;
- investors' receiving a possibility to charge for using the crated objects of transport and logistical infrastructure.

Within the last item, it is necessary to emphasize that paid transport and logistical objects should have free alternative—in order to avoid creation of natural monopolies.

Thirdly, it is necessary to stimulate growth of consumers' trust to Internet deals. It is the most difficult task as it is related to transformation of public opinion and fighting natural opposition to changes in society. This supposes realization of the following basic measures:

- conduct of a series of open consultations for wide groups of population regarding possibilities and advantages of conclusion of Internet deals;
- propaganda of Internet deals with the help of social;
- providing Internet entrepreneurs with 3% corporate tax exemption for provision of pricing advantage (temporary measure).

Realization of the offered recommendations will allow quickening the process of formation of Internet economy in modern Russia.

References

Bell D (1973) The coming of post-industrial society: a venture in social forecasting. Basic Books, New York

Birner J, Garrouste P (2013) Markets, information and communication: Austrian perspectives on the internet economy. Taylor and Francis, Abingdon, pp 1–320

C-News (2017) E-commerce market in Russia reached $13.9 billion. http://www.cnews.ru/news/top/2016-03-02_rynok_elektronnoj_kommertsii_v_rossii_dostig_13. Accessed 8 Jan 2017

Chen Y (2016) Financial system innovation of large-scale LED enterprises in China within the internet economy. Light Eng 24(3):132–135

Choi SB, Williams C, Ha SH (2014) Institutions and broadband internet diffusion in emerging economies: lessons from Korea and China. Innov Manage Policy Pract 16(1):2–18. doi:10.5172/impp.2014.16.1.02

Dubrovskaya T, Kozhemyakin E (2017) Media construction of Russia's international relations: specifics of representations. Crit Discourse Stud 14(1):90–107. doi:10.1080/17405904.2016.1196228

Dudina VI, Judina DI, King EJ (2017) Fears about antiretroviral therapy among users of the internet forum for people living with HIV/AIDS in Russia. AIDS Care Psychol Socio-Med Aspects AIDS/HIV 29(2):268–270. doi:10.1080/09540121.2016.1211607

FOMnibus (2016) Number of Internet users in Russia. http://www.bizhit.ru/index/users_count/0-151. Accessed 8 Jan 2017

Fuchs C (2009) Information and communication technologies and society: a contribution to the critique of the political economy of the internet. Eur J Commun 24(1):69–87. doi:10.1177/0267323108098947

Fuchs C, Bolin G (2012) Introduction to the special section "critical theory and political economy of the internet (nordmedia 2011)". TripleC 10(1):30–32

Gilder GF (1979) The make-work economy. McGraw-Hill, New York

Gnezdova YV, Chernyavskaya YA, Rubtsova LN, Soldatova NF, Idilov II (2016) Modern aspects of the development of internet-economy in Russia. J Internet Bank Commer 21 (Spec.Issue 4):128–131

Government of the Russian Federation (2014) Decree of the Government of the Russian Federation No. 313 dated April 15, 2014. On establishment of the state program of the Russian Federation "Information society" (2011–2020). http://pravo.gov.ru/proxy/ips/?docbody=&nd=102349623&rdk=&backlink=1. Accessed 8 Jan 2017

InSales (2015) Analytical bulletin InSales 2015: market of internet commerce in Russia in 2014. http://www.insales.ru/blog/2015/05/20/analytical-bulletin-insales-2015. Accessed 8 Jan 2017

Janssen E (2008) What a lie! Esquire. http://esquire.ru/economic-bubble. Accessed 8 Jan 2017

Jarrett K, Wittkower DE (2016) Economies of the internet. First Monday 21(10):114–119. doi:10.5210/fm.v21i10.6939

Lavrinenko O, Ohotina A (2015) The influence of enterprises' economic activity integration into internet environment on the economy. Actual Prob Econ 166(4):436–446

A.P. Sukhodolov et al., *Internet Economy vs Classic Economy: Struggle of Contradictions*, Studies in Computational Intelligence 714,
DOI 10.1007/978-3-319-60273-8

Maksiyanova TV (2013) Place and role of Internet economy in economic science. Issues Mod Sci Pract 1(45):162–168

Marcus D (2011) Prospective analysis on trends in Cybercrime from 2011 to 2020. https://www.mcafee.com/us/resources/white-papers/wp-trends-in-cybercrime-2011-2020.pdf. Accessed 8 Jan 2017

Marshall A (1923) Money, credit and commerce. Amherst, New York

Metcalfe B (2013) Metcalfe's law after 40 years of Ethernet. IEEE Computer, New York

Ministry of Economic Development of the Russian Federation (2010) Decree of the Ministry of economic development of the Russian Federation No. 54 dated February 15, 2010. On the order of public auction in electronic form during sales of property (company) of debtors in the course of procedures applied in the case on bankruptcy, Requirements to electronic platforms and operators of electronic platforms during public action in electronic form during selling the property (company) of debtors in the course of the procedures applied in bankruptcy cases, and the order of confirmation of correspondence of electronic platforms and operators of electronic platforms. http://economy.gov.ru/minec/activity/sections/corpmanagment/bankruptcy/elground/doc20100215_020. Accessed 8 Jan 2017

Ministry of Internal Affairs of the RF. The Main Information and Analytical Center (2017) The state of crime in Russia in 2011–2016. https://мвд.рф/reports/7/. Accessed 8 Jan 2017

OECD (2013) The internet economy on the rise: progress since the Seoul declaration. Organisation for Economic Cooperation and Development (OECD), Paris, 9789264201545, pp 1–180

Powers SM, Jablonski M (2015) The real cyber war: the political economy of internet freedom. University of Illinois Press, Illinois, pp 1–274

PricewaterhouseCoopers (2015) Survey of customers of Internet stores. http://www.marketing.spb.ru/mr/social/retail_ecomm.htm. Accessed 8 Jan 2017

Sakaiya T (1991) The knowledge-value revolution, or, a history of the future hardcover. Kodansha America, New York

Sakoyan A (2014) Internet assets for high price. http://polit.ru/article/2014/08/21/as200814. Accessed 8 Jan 2017

Schumpeter JA (1939) Business cycles. McGraw-Hill, New York

Simon G (1993) Rationality as a process and product of thinking. Thesis 3(1):16–38

Sokolov M (2016) Russian youth shows interest in IT. https://rg.ru/2016/02/18/anna-kulashova-rossii-nuzhno-kak-minimum-udvoit-kolichestvo-programmistov.html. Accessed 8 Jan 2017

Sombart W (1930) Die drei Nationalökonomien, München & Leipzig. Duncker & Humblot, Berlin

State Duma of the RF (2000) Project of the Federal Law "On e-commerce" No. 11081-3 dated October 3, 2000. http://www.ekey.ru/info_def/legally_concerned/e_trade/copy_of_0. Accessed 8 Jan 2017

State Duma of the RF (2005) Federal law No. 94 dated July 21, 2005, On г. On placement of orders for supply of products, execution of works, and provision of services for state and municipal needs. http://www.consultant.ru//cons_doc_LAW_54598. Accessed 8 Jan 2017

State Duma of the RF (2013) Federal law of the RF dated April 5, 2013, No. 44. On the contract system in the sphere of purchases of products, works, and services for provision of state and municipal needs. http://www.consultant.ru//cons_doc_LAW_144624. Accessed 8 Jan 2017

Stewart TA (1997) Intellectual capital. The New Wealth of Organizations, Doubleday, New York

TASS (2017) Share of non-cash payments in the RF will grow by 20% over 5 years. http://tass.ru/ekonomika/3606247. Accessed 8 Jan 2017

Taylor NG, Jaeger PT, McDermott AJ, Kodama CM, Bertot JC (2012) Public libraries in the new economy: twenty-first-century skills, the internet, and community needs. Publ Libr Q 31 (3):191–219. doi:10.1080/01616846.2012.707106

Valtukh KK (2001) Information theory of cost and laws of non-equilibrium economy. Janus-K, Moscow

Vendil Pallin C (2017) Internet control through ownership: the case of Russia. Post-Sov Aff 33 (1):16–33. doi:10.1080/1060586X.2015.1121712

Yadav R, Chauhan V, Pathak GS (2015) Intention to adopt internet banking in an emerging economy: a perspective of Indian youth. Mark Intell Plan 33(4):530–544. doi:10.1108/IJBM-06-2014-0075

Zubraeva (2015) Customers are drawn into the network: internet commerce ousts traditional stores. Russian Business Newspaper No. 983 (4). https://rg.ru/2015/02/03/torgovlya.html. Accessed 8 Jan 2017

Printed in the United States
by Baker & Taylor

Printed in the United States
By Bookmasters